LIFE UNFOLDING

Jamie A. Davies is Professor of Experimental Anatomy at the University of Edinburgh. He has published over 100 research papers in the field of mammalian development, authored *Mechanisms of Morphogenesis* (Academic Press, 2nd edition 2013), and edited three multi-author books in the fields of development, stem cells and tissue engineering. He is a Fellow of the Society of Biology, a Fellow of the Royal Society of Medicine, and the Royal Society of Arts.

Life Unfolding

How the human body creates itself

JAMIE A. DAVIES

OXFORD
UNIVERSITY PRESS

OXFORD

UNIVERSITY PRESS

Great Clarendon Street, Oxford, OX2 6DP,
United Kingdom

Oxford University Press is a department of the University of Oxford.
It furthers the University's objective of excellence in research, scholarship,
and education by publishing worldwide. Oxford is a registered trade mark of
Oxford University Press in the UK and in certain other countries

First published 2014
First published in paperback 2015

Impression: 1

Published in the United States of America by Oxford University Press
198 Madison Avenue, New York, NY 10016, United States of America

British Library Cataloguing in Publication Data

Data available

Library of Congress Cataloguing in Publication Data

Data available

ISBN 978–0–19–967353–7 (Hbk.)
ISBN 978–0–19–967354–4 (Pbk.)

Printed and bound in Great Britain by
Clays Ltd, St Ives plc

To Katie

CONTENTS

CONTENTS

ACKNOWLEDGEMENTS

I would like to thank Dr Katie Brooks for her encouragement, for her patience when writing this book had made me lose track of the time *yet again*, and for her very helpful comments on the first draft.

I would also like to thank the following colleagues in the field of developmental biology and related disciplines, who did me the enormous service of checking sections of this book in which they have world-leading expertise. They are James Briscoe, Mike Clinton, Kim Dale, Megan Davey, Peter Kind, Val Wilson, Georgia Perona-Wright, Thomas Theil, and Cheryl Tickle. For any errors that remain, the blame is entirely mine. Finally, I would like to thank Latha Menon and her colleagues at Oxford University Press for their immensely helpful editorial advice.

The title *Life Unfolding* is inspired by a phrase from Gillian K. Ferguson's biological poem, *Not in charge*.

ETHICAL STATEMENT

This book, which describes the mechanisms of human development, includes published knowledge that has been obtained by study of human embryonic and foetal material and by experimentation on living animals. Because academic publishers and funders of research require that all work be ethically approved by the relevant independent ethics committees, I have assumed that these experiments were conducted according to the standards of the day. Ethical standards do change with time, and some work done long ago would not be permitted now. Mention of experimental results in this book does not imply personal ethical endorsement of the experimental method, by either the author or publisher.

A NOTE ON REFERENCES AND FOOTNOTES

This book is aimed at a general readership and, for this reason, the mechanisms of human development have been described with as little molecular detail as is possible. Long lists of unpronounceable protein names are tedious even to professional biologists and would be out of place in a book that is meant to present deep principles. Nevertheless, to help committed students of biology and medicine to connect the events in this book with papers on molecular embryology, occasional technical footnotes indicate which details the main text is avoiding. Understanding these technical notes is not necessary for understanding the book—they are present only for a small subset of the readership with particular needs.

For similar reasons, the main text uses superscript numbers to cite technical research references, especially in support of statements that are at variance with a traditional view. In the interests of conciseness, these citations have been kept to a minimum and, instead of following the normal academic convention of citing original experimental reports, they instead often point to review articles that provide a convenient gateway to experimental literature. No explicit citations are given for material that is well covered in standard textbooks. The alternative, of having the text interrupted by thousands of citation numbers, as would happen in a research monograph, would not be appropriate here.

A list of accessible books for further reading on particular topics is given at the end of the book.

INTRODUCTION

The possession of knowledge does not kill
the sense of wonder and mystery.
There is always more mystery.

Anaïs Nin

1

CONFRONTING AN ALIEN TECHNOLOGY

The history of a man for the nine months preceding his birth
would probably be far more interesting than all the
three-score and ten years that follow it.
                                        Samuel Taylor Coleridge

When the English philosopher-poet Coleridge wrote those words, he was expressing, in the elegant language of an adult, the wonder that is felt by every child who asks his parents *'How did I get here?'*. To many parents, the question brings up potentially uncomfortable issues about sex and about when, and how much, a child should be told. To the child asking, innocent of these adult psychosocial complexities, the question is both simpler and far more profound: it is simply about how a new person can come to be.

No child has ever received a full and correct answer to the question, because nobody yet knows enough to give one. When Coleridge was writing, some facts were known about the sequence of anatomical changes that took place as a new human grew in the womb, but how or why they took place remained utterly mysterious. In the two centuries that have since passed, generations of scientists have laboured to understand how a fertilized egg can become a child. Research has pushed forwards quickly in the last decade and, as more intricate and complex mechanisms have been discovered and the mystery has decreased a little, the collective sense of awe has only increased. The story that is being unearthed, and presented so far mainly in the dry text of learned journals, is an astonishing one. It is the story of something every one of us has done, and it is therefore a story that belongs to us all. This book is an attempt, by a scientist who has the good fortune to be working in this area, to bring together highlights of recent research into that most profound, and childlike, of questions—*how did I get here?*

Our current understanding of human embryonic development has not come from a single investigative approach, but is instead emerging from a synthesis of information obtained from a range of scientific disciplines. Embryology and neonatology are concerned directly with development and produce a great deal of directly relevant anatomical and functional information. Genetics and toxicology, both fields with a scope much broader than development, have been very valuable in identifying precise causes of congenital abnormality: this is important because known causes of malformation can point developmental biologists to identify the molecular pathways that are required for the affected part of the body to grow normally. Biochemistry and molecular biology are invaluable for working out the details of how these pathways work, following the logic of development even down to the spatial scale of interactions between the atoms of biological molecules. Cell biology accounts for how molecular pathways are brought together to control the behaviour of individual cells. At much larger spatial scales, disciplines such as physiology, immunology, and neurobiology uncover the ways in which multitudes of cells communicate and coordinate.

All of the subjects mentioned so far would be found in departments of biology or medicine, the traditional homes of embryological work. In recent years, however, exceptionally valuable insights into human development have also been contributed by researchers in fields that might seem at first to have nothing to do with the topic, such as mathematics, physics, computer science, and even philosophy. These efforts have not been focussed on precise details of which cell does what and when, but have instead tackled the profound, abstract questions that development raises, including, *how can the simple become complex?*, *how can error-prone mechanisms construct something precise?*, and *is human development too complex for developed humans to understand completely?* The jury is out on the last of these questions—the word 'completely' being the point of dispute—but significant progress has been made on the other two questions, both of which find an answer in the related concepts of 'emergence' and of 'adaptive self-organization'. The terms are essentially two sides of the same thing, one viewed from above and one below. 'Emergence' tends to be used by people looking down from the perspective of high-level behaviours, and is the process by which complex structures and behaviours arise from simple components and rules. 'Adaptive self-organization' is a description grounded in the components and looks upwards, describing how the application of simple rules to these components can result in their collectively doing something large scale, clever, and subtle.* The way in which adaptive self-organization allows non-living molecules to produce a living cell, and allows cells with very limited individual abilities to produce a very able multicellular body, will form a theme that runs through all of this book because it is the core of development. Adaptive self-organization and emergence go far beyond biology, and some very readable books on its wider implications are listed under 'Further Reading'.

One very clear message that can be taken from our emerging understanding of development is that the self-construction of a body is very different from our normal notions of construction, experienced in architecture or engineering. This highlights a fact that is as important as it is ironic: the methods by which we build our own bodies seem utterly alien to us. It may therefore be useful, in preparing to understand how an embryo can build itself, to compare and contrast the biological system with human-scale construction technologies.

Engineering projects, for example the construction of a locomotive or a building, tend to share a common set of features. To begin with, there is a specific plan, set out in a blueprint or some other form of schematic, that shows a clear representation of the intended outcome. The plan depicts the finished structure but will never itself be a physical part of it. Each project has someone in overall charge, a chief engineer or an architect, and he or she operates through a hierarchy of command to pass instructions to the artisans who carry out the work of cutting, brick-laying, welding, and painting. The components that are handled by these artisans cannot pull themselves together on their own. Instead, they have to be cemented or bolted or welded by workers who remain distinct from the structure. These workers, and their chief engineer, provide vast amounts of 'outside' information, such as knowledge of the techniques for soldering or for shaping an arch, information that is not present in what they are building. Finally, most structures have to function only when they are complete.

---

* Synonyms for, or sub-types of, adaptive self-organization include 'swarm intelligence' and 'hive mind'. These are often used in studies of social insects or even populations of humans but use words that seem too suggestive of sentience to be applied to mere chemicals and cells. For this reason, in previous books and in this one, I have used the term 'adaptive self-organization', used more commonly in the physics and mathematical communities to refer to the same phenomenon.

Searching for any of these features in biological construction highlights just how different life is from conventional mechanical and civil engineering. Unlike engineering projects, biological construction does not involve a blueprint-like representation of the final structure. There is certainly information in a fertilized egg (in genes, in molecular structures, and in spatial variations in the concentrations of particular chemicals), but this has no simple relationship with how the final built body will look. We know that the information has the effect of controlling the sequence of events that will follow (we know because changing the information, for example by mutating a gene or by altering the location of a relevant chemical, alters that sequence and makes development abnormal).

In engineering, and especially in mathematics, a final form can be specified by a series of instructions. Telling someone to drive a peg into the ground in the middle of a wheat field, tie a rope to it, hold the other end, walk away until the rope is fully tight, then turn right and walk along so that the rope remains tight is, for example, a way of specifying the construction of a simple crop circle. Some structures can be specified much more economically by the use of instructions than by producing detailed blueprints. If you happen to have a pen and paper handy, try following these instructions to draw a geometrical figure called a Sierpinski Gasket:

1. Draw an equilateral triangle, with a horizontal base, as large as possible. Consider this the 'triangle of interest'.
2. Draw three lines inside the triangle of interest, each line going from the mid-point of one side to the mid-point of an adjacent side. These three lines will have defined a downward-pointing triangle occupying a quarter of the area of the triangle of interest.
3. Shade in the triangle you just created.
4. Observe that there are now three unshaded upward-pointing triangles within the old triangle of interest. Consider each of these a triangle of interest and go back to step 2 for each of them.
5. (Stop when you get bored: if you had a fine enough pencil, this would go on for ever.)

The Sierpinski Gasket, which would be gasket-like if the shaded areas are imagined as holes, is an example of a fractal, or self-similar, structure whose form looks similar at any magnification. Another is Cantor Dust, which can be drawn most easily on a medium that allows erasure, such as a blackboard: draw a line, erase its middle third, then erase the middle thirds of the shorter lines this erasure created, and so on. After a while, you will be left with a peculiarly spaced set of chalk marks, the intervals between which have statistical properties identical to those of a vast range of natural phenomena, such as the distribution of sizes of avalanches in a sand dune, intervals of time between drips of a leaky tap, or intervals between large earthquakes, epidemics, and major extinctions.

Even outside mathematics, the idea of specifying a constructed object by giving the rules to generate it, rather than by depicting a detailed illustration of its final form, is common. Food recipes work this way, as does textile manufacture, at levels ranging from the simple 'knit one, purl one' type of instruction set in a knitting pattern to the complicated punched cards of Jacquard's 1801 loom, the world's first programmable manufacturing robot. Music is similarly specified by instructions, in this case by dots on a stave that instruct a musician to play a certain pitch, at a certain time, for a certain duration.

Our long cultural experience with the use of instructions as an economical method of specifying an intended outcome makes it dangerously natural to accept that biological information specifies our own form by a similar process to those mentioned above. There is one critical difference: human-built objects constructed by following a sequence of instructions are read and acted on by an external, intelligent agent. Even apparent exceptions to this statement, such as a knitting machine or a player piano, rely on machines that are themselves made from plans or instructions that were read by external, intelligent agents, so they are not really exceptions. Put simply, cardigans, symphonies, cars, and cathedrals do not build themselves. Instructions, operational knowledge (the skills of knitting, cooking, welding, stone-masonry, etc.), and physical manipulation of materials all come from outside, not from within the growing structure. The information in an embryo, on the other hand, has to be read and acted upon by that embryo, with no handy external workmen to do the heavy lifting or the heavy thinking. As will become clear shortly, this means that responsibility for biological construction is shared between all of the components involved rather than, as in most technological construction, someone being in overall charge. In constructing a human being, control is not exercised by a few privileged parts but emerges from the system as a whole.

Understanding any construction process demands, in addition to other things, some appreciation of the nature of the materials to be used. Near my laboratory in the University of Edinburgh, there are three famous bridges: Thomas Telford's elegant Dean Bridge in the city itself, Benjamin Baker's iconic

Forth Railway Bridge across an inlet of the sea, and the Forth Road Bridge that now runs beside it. Telford built his bridge from stone blocks: heavy, bulky components that are safe in compression only. He therefore used the conventional method of first building the pillars, then making a wooden scaffold to set out the arch, then covering it with shaped arch stones and building up until the weight of the stone stabilized the arch and the scaffold could be removed. Baker used what was then a radical new material—steel—to construct his rail bridge. This material can be used in tension and compression, so the bridge could be built outwards from each set of piers as a cantilevered structure, cranes being used to hoist the long and comparatively light steel sections in place and rivets being used to connect them together. The suspension bridge that carries the road, the newest of the three bridges, hangs from tense steel cables that rise up to, and press down on, towers on each shore. For this bridge, the towers had to be built first, then strong anchor points were made behind them for the cables, and then the cables themselves were added strand by strand and tensioned until they were complete and the road could be hung from them. In each case, the entire strategy for building the bridge was determined by the nature of the materials, and none of the bridges could have been constructed using the strategy meant for either of the others. In biology too, the strategy for construction depends on the nature of the components involved. This is therefore a good place to introduce three key biological components that will be mentioned many times in this book: proteins, mRNA, and DNA.

By far the most important molecules in biological construction are the proteins. They make most of the physical structures that give cells their shape, they form the channels and pumps that regulate what enters and leaves the cell, and they are the catalysts that drive and control the chemical reactions of life. These reactions include the metabolic pathways that make the body's other components such as DNA, fats, and carbohydrates. The relative importance of proteins is illustrated by the fact that red blood cells naturally throw away their nuclei, containing all of their genes, during maturation, yet they continue to live for around a hundred and twenty days after doing so. A cell that kept its genes but lost the function of its proteins would be dead within seconds.

A protein consists of a long chain of individual units, amino acids. There are twenty different types of amino acid, and they vary in shape and chemical properties. They interact with one another, and this means that chains of amino acids tend to fold into complex forms, either on their own or with the temporary help of other proteins. This folding process is so involved that it is still not possible to deduce, mathematically, the final form of a protein simply from knowledge of its amino acid sequence. (Computer programs for predicting protein shapes do exist, but they use a combination of calculation and probabilistic reasoning based on the relationships between the known structures of other proteins, determined the hard way by X-ray crystallography, and their underlying amino acid sequences. They therefore operate in a manner similar to computer programs that perform weather forecasting, albeit with a little more success.)

Different proteins have different sequences of amino acids. These are added, one by one, to the growing chain of a protein as it is being made in an order that is specified by another molecule, called messenger RNA or mRNA (see Figure 1). Molecules of mRNA also consist of single chains of individual units, the RNA

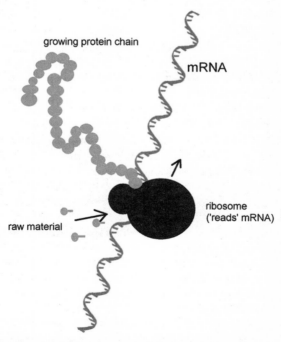

growing protein chain

mRNA

ribosome
('reads' mRNA)

raw material

FIGURE 1 mRNA is translated into a protein by a ribosome, which assembles amino acids into a growing protein chain according to the sequence of bases in the mRNA.

bases A, C, G, and U. These are structurally similar to each other and, compared to amino acids, chemically rather dull: molecules of mRNA do not do much in the cell except for directing the sequence of amino acids to be added to a growing protein. This sequence is determined by the sequence of bases in the mRNA, a group of three bases specifying each amino acid.

The sequences of bases in molecules of mRNA are determined directly by the sequences of DNA bases. DNA is a very long molecule consisting of a succession of four bases, A, C, G, and T, which can be accommodated in any sequence. The individual DNA molecules that form the cores of the forty-six chromosomes in each of our cells are many millions of bases long. Within these are the stretches that constitute individual genes. When a gene is being read, an RNA molecule is made such that it copies, in the language of the RNA bases A, C, G, and U, the order of the DNA bases A, C, G, and T. The RNA is therefore effectively a copy, or 'transcript', of the gene in a different medium. The actual reading of genes is performed by complexes of proteins. These first bind to various short sequences of bases such as ATAAT, or TCACGCTTGA, which can be found near the start of a gene. Different genes have different combinations of these short sequences near them, and each sequence binds its own particular protein, so different combinations of proteins are involved in activating the reading of different genes.

The fact that different genes are activated by different DNA-binding proteins is important because different cells of the body need to make different types of protein. Cells in the gut, for example, make proteins that digest food, cells in the ovary express proteins that make reproductive hormones, and white blood cells make proteins that fight infection. All of these still contain all of the genes of the genome, even those that they never use. Only the genes that the cell needs are read, because only the set of DNA-binding proteins that recognizes exactly those genes is present.

This is where we are forced to abandon the idea of any one of these components being in overall charge of a cell, or of an embryo. To recap: proteins are made only because active genes specify (via mRNA) that they should be made. Those genes are active only because proteins already present make them so. The logic is therefore circular: control is located nowhere, because control is located everywhere (Figure 2).

The circularity of Figure 2 has an interesting implication. For a cell to remain in a stable state, the total set of active genes must include those that specify the proteins that bind to recognition sequences near those very genes, but must *not*

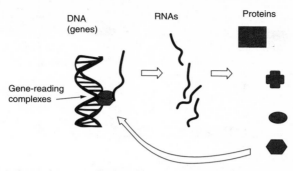

FIGURE 2 The circular nature of biological logic. Proteins decide which genes are read, and these genes specify the production of new proteins, some of which decide which genes are read...and so on.

include any proteins that will activate currently inactive genes. Unless these conditions are met, the proteins made by the set of genes active now will not sustain that set of genes—some genes may go off and others on, and a new set of proteins will be made, and so on. These changes will continue to take place until a self-sustaining state is reached. This is the basis of how, as we develop, some of our cells change to become new cell types. Such change is driven typically by external influences, 'signals', that alter the ability of specific proteins to activate genes: these disrupt what was an existing stable state, and make it shift to a new state. Many examples of such signals will feature in the rest of this book.

Distributed, circular control is by no means the only strange feature of biological construction. Another feature that seems very alien, when viewed from the comfort zone of conventional engineering, is that biological molecules can assemble themselves into larger-scale structures spontaneously, something that bricks and bolts never do. This process, which is of fundamental importance for the existence of life, is a little like the formation of crystals. Ordinary crystals, such as those grown by children with chemistry sets, form because their constituent molecules can bind to each other, typically by attractions of small, local electric charges. Proteins also have patterns of intrinsic local electric charges, often in rather complex crevices in or projections from the main body of the protein. The arrangement of charges and the shape of the protein are properties that derive from the sequence of amino acids. Sometimes, a protein has one kind of crevice at its front end, and a projection that fits that type of crevice at its back end, rather like a 'Lego' brick. In that case, molecules of this

11

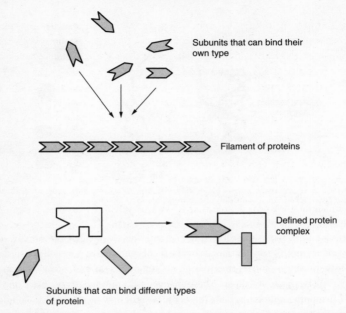

FIGURE 3 Proteins often carry charge patterns, projections, or crevices that can bind strongly to complementary charge patterns or shapes on other proteins. Where one end of a particular protein can bind the type of structure at the other end of the same protein, then molecules of this protein can form a filament, with each molecule behaving as a separate link in a chain. Where a particular protein can bind only proteins of a different sort, multi-protein complexes of a defined size and shape are created. The protein complexes that transcribe genes are of this type.

protein can line up end to end to make a long, thin filament of indefinite length (Figure 3). More often, each protein can recognize binding sites only on another specific protein, or some other molecule, and not on itself. This means that it cannot form indefinite crystal-like threads with identical molecules, but instead binds only to a defined number of other proteins to make a multi-component complex of a defined structure. These complexes are very important in the cell because they act as tiny machines that can run complex chemical reactions or organize the assembly of structures that are too large and complicated to arrange themselves spontaneously. The gene-reading protein complexes already mentioned are an example.

The level of organization represented by protein complexes takes us to a very important boundary. The assembly of proteins into their complexes relies on information that resides only within proteins themselves ('information' being, in this case, synonymous with structure). It therefore belongs within the domain of chemistry and the result is always the same: reliable, reproducible, but inflexible. At larger scales, biological structures are more variable, their exact arrangements being adapted to circumstances. The overall shape of a cell, for example, is adapted to the space it must fill in a tissue. The arrangement of the connections it makes with neighbouring cells must similarly be adapted to the location of the cells that surround it. These larger-scale structures cannot, therefore, be determined solely by the information contained in the chemical structure of their molecular components: extra information is needed. This transition, between internally determined structure and structure that is regulated by external information as well, takes us across a boundary from pure chemistry to the realm of biology. In biological systems, layers of regulation are added to chemical self-assembly to produce systems that organize structures adapted to circumstance and need. This is where the concept of adaptive self-organization, mentioned earlier, becomes important. Adaptive self-organization turns out to be the key to explaining how a few thousand genes and proteins, none of which can possibly hold any concept, in any language, of the structure and function of a human body, can nevertheless organize themselves to build one. It stands in marked contrast to the way that engineering projects use external agents, such as workmen or robots, to assemble components together in the right way. The following chapters will illustrate how adaptive self-organization is critical to human development, at levels ranging from self-organization of molecules within a single cell to the large-scale construction of complex tissues.

A final peculiar aspect of biological construction comes from a great restriction of life: it cannot be stopped and re-started to suit the needs of the building process. Human-engineered structures, such as computers and aeroplanes, are expected to function only when they have been completed: there is no need for a partially complete structure to do anything useful. Development of an embryo has the constraint that every stage of development must be compatible with its staying alive. A plumber wanting to connect a new branch to the water main of a building can shut the water off, cut through the main pipe, add a 'T' piece, then turn the water on again when the job is completed. If a growing human were to take the same approach when it needed a new branch from a main artery, it would have bled to death well before the job could be completed. The same is true for all of the other essential systems of the body. The absolute requirement for constant viability in the face of developmental change is a very serious constraint on how bodies can be built, and it is another reason that the construction of a body can seem so alien, and sometimes so complicated, when compared to ordinary engineering.

When we seek to understand our own beginnings, we must be prepared to move beyond homely analogies, based on how we build things, and see the embryo in its own terms. It will be a journey into strange territory, which demands new ways of thinking and a letting go of engineering metaphors. We do not build embryos, after all: they build us.

PART I

FIRST SKETCH

2

FROM ONE CELL TO MANY

I am large, I contain multitudes.   Walt Whitman

It is one of the great ironies of biology that the human body, arguably the most complex single entity in the known universe,* develops from a very simple beginning. An adult human consists of over a million million cells—a number that is about ten times the number of stars in our galaxy or, to give a more down-to-earth comparison, about ten times the number of grains of sand on a beach volleyball court. These million million cells are not heaped up haphazardly but are located and connected in patterns so intricate that, even after two millennia of research into anatomy, we have still not worked out all of the details. There are hundreds of different cell types amongst them, each with its own function and way of living, each produced and renewed in the correct proportions and the correct places. All of this elaboration originates from one single, simple, almost-featureless cell, the fertilized egg. It is from this unassuming beginning that the complexity of a human being has to create itself—to pull itself up by its own bootstraps, as the saying goes.

The first major step towards a more complex form of being is the conversion of that single cell into a multitude. This is necessary because any complex living animal requires many different things to be going on at the same time. Right now, you are breathing air, digesting food, detoxifying chemicals, growing hair,

---

* Humans are singled out on the assumption that the wiring of our brains includes more underlying neural complexity than is present in other mammals, although future research may prove this assumption wrong.

making new skin cells, filtering blood, fighting would-be invaders, regulating temperature, hearing, reading, thinking and, by this stage of the sentence, probably indulging in some introspection. These activities, and hundreds of others not mentioned but happening anyway, use different sets of proteins and biochemical pathways. Many would be downright incompatible if they were to take place in the same place: consider, for example, a mother's making milk for her baby and her digesting milk she has just drunk in her tea. There are many other examples that are incompatible for more subtle reasons concerning the details of protein and gene function.

Complex organisms deal with this problem using compartmentalization, the principle of separating each activity into different places. Bodies are divided into organs that do different things, and organs are divided in their turn into tissues, which perform different functions of the organ. Tissues are divided into cells, different types of cells again performing their own specific tasks. Within each cell, though, most molecules can move around freely and it is difficult for many different things to take place at once. There are some internal compartments to cells, and the ability of different parts to perform slightly different functions will be a central theme in Chapter 8, which describes how cells move and navigate around an embryo. Even so, the ability to perform several tasks at once is limited, and the cell can therefore be considered as a basic unit that does only one or two things at a time. For this reason, having a multiplicity of different cells is an essential step in making a complex body.

The mechanisms by which one cell can become two and, by repeating this, can become many, are not only critically important to embryonic development: they also illustrate clearly how small, simple molecules can organize themselves to achieve remarkable feats at scales far larger than themselves, and how they can build structures of amazing detail with no prior plans. This idea is central to understanding the embryo as a whole. This chapter will therefore be devoted to the mechanisms of cell division, which we can then take for granted in all of the later chapters.

The fertilized egg with which human development begins is unusually large, about a tenth of a millimetre across and just visible to the naked eye. Most cells in the mature body are very much smaller, being about a hundredth of a millimetre across and having about a thousandth of the volume of the egg. This means that the fertilized egg can generate a many-celled embryo simply by dividing itself up, first into two, then four, then eight and so on, without the need to pause for growth. This form of cell multiplication, called cleavage, is very useful because it means that the business of getting food to fuel growth can be deferred until the embryo is already multi-cellular and therefore able to dedicate specific parts of itself to procuring food.

With no growth taking place, dividing one cell into two is mostly a case of sharing out all of the internal molecules, such as proteins, equally between the daughter cells. The very nature of division with no change in net volume means that the concentration of internal proteins and nutrients is unchanged. The glaring exception to this general statement is DNA: the undivided cell has forty-six chromosomes (twenty-three from the embryo's mother and twenty-three from the father), and each of the cells produced will also need forty-six chromosomes. The chromosomes therefore have to be replicated before each round of cell division begins. What is more, some system must exist to ensure that the replicated chromosomes are allocated equitably to the daughter cells, not just so that each daughter cell receives a total of forty-six, but to ensure specifically that each receives exactly one copy of each chromosome that was inherited from the father and exactly one copy that was inherited from the mother. The system that achieves this accurate separation of chromosomes is a central feature of animal and plant life and it has existed for around 2,500,000,000 years. Only for the last two million years or so has it been producing an animal capable of starting to understand it.

The copying of the DNA is in many ways the simplest part of the process, and is also the oldest, having existed in a basic form for at least 3,500,000,000 years. It uses the fact that DNA molecules exist as a pair of nucleotide chains (sometimes called 'strands'). Where there is an 'A' nucleotide on one chain, there is always a 'T' facing it on the other, and where there is a 'C' on one chain, there is always a 'G' facing it on the other. This rigid pairing rule, which arises simply from the detailed chemical shapes of the nucleotides A, T, C, and G, means that each single chain of DNA carries enough information for the sequence of its partner chain to be deduced. When a cell needs to replicate DNA, a complex of enzymes first separates the two chains. It then assembles a new partner chain for each of the originals, bringing new nucleotides together in the order determined by the order of nucleotides on each original chain. Each new chain stays with the old one that was used to specify its construction, and the result is two DNA double-chained molecules where there used to be one. The DNA has, effectively, been copied. The proteins of the chromosome, around which DNA wraps, are added once the DNA has been copied.

Once the forty-six chromosomes of the single-celled embryo cell have been replicated, they have to be moved so that precisely one copy of each chromosome from the mother and one copy of each chromosome from the father end up in each daughter cell. This task can be broken down into a number of sub-tasks: (1) defining where the centres of the two daughter cells will be; (2) lining up all of the copied chromosomes exactly between these centres; (3) pulling the copies apart so that one copy of each pair goes to each daughter; and (4) separating the daughter cells. Each of these events involves coordinated actions at scales vastly larger than the sizes of the individual molecules that drive them, and each has to take place properly even though the precise prior location of key components, such as chromosomes, will be quite variable. The events therefore rely heavily on adaptive self-organization and provide an excellent example to illustrate this principle.

The first problem is that of defining the centres of the new daughter cells. It is easiest to start by understanding how a normal adult cell—one that is not preparing to divide but just resting—defines its centre. The problem seems trivial at first meeting, but the more it is considered, the tougher it becomes. Cells in general are not of an exactly predictable shape, and the shapes of many depend closely on their immediate environment: this rules out using any kind of pre-determined plan. The diameter of a typical human cell, about a hundredth of a millimetre across, seems tiny to us, who consist of a million million cells; it is, however, a thousand times larger than the length of a typical protein molecule. Yet complexes of proteins have somehow to find the cell centre. To represent the problem at a whole-body scale, it is like individual humans who are blind-folded and deaf, and can communicate only by touch, being placed inside the Albert Hall and having to cooperate to find its middle.

The mechanism used is ingenious and illustrates how apparently trivial details of biochemistry can be critically important to the functioning of a cellular machine. The star of the show is the protein tubulin, molecules of which can associate with each other to form long tubes called microtubules. A quirk of the method by which individual tubulin molecules associate means that the process of a few tubulin molecules getting together to begin a new microtubule is a very improbable event, whereas the process of a tubulin molecule joining on to an existing microtubule to extend it is comparatively easy. Microtubules, therefore, tend not to form spontaneously although, once formed, they do tend to grow. A second quirk of tubulin biochemistry means that each individual tubulin molecule can exist in one of two states, referred to here as fresh or stale,[†] and fresh molecules slowly decay into stale ones. Only fresh molecules join on to the end of existing microtubules. Microtubule ends are stable only if they are made of tubulin in its fresh state (but it does not matter if the tubulins along the main length of a microtubule go stale, as long as the end is still fresh).[1] If the end goes stale, it starts to unravel, and the unravelling will continue along the microtubule until it reaches some stable tubulin in its fresh state. Given that tubulins in the main length, back from the end, must have been in the microtubule for even longer than those that went stale at the end, they will probably have gone stale long ago and there will probably be no fresh tubulins left to stop the unravelling. The microtubule will therefore fail, catastrophically. The only way for a microtubule to be protected from unravelling, without special help from other molecules, is to grow quickly enough that it adds fresh tubulins faster than they can decay. Left to themselves, microtubules are therefore generally either growing quickly or collapsing catastrophically, and the constant probability of entering the collapsing state means that there are always fewer long tubules than short ones. This is very important to the mechanism that cells use to find their centres.

Because tubulin molecules rarely come together to make new microtubules spontaneously, the cell includes special complexes of proteins that can catalyze the process. These complexes are located in one key place, the centrosome, and microtubules radiate away from this place like spokes from a hub.[2] As long as they grow quickly enough that their ends consist of tubulin in its fresh state, these microtubules will keep extending towards the edge of the cell. There are two theories about how they help the centrosome to move to the cell's centre, and each theory has been supported by experimental evidence obtained in different organisms. It is not yet clear whether one or the other, or perhaps both, are used in human embryos. One depends on pushing and the other on pulling.

The pushing mechanism[3] uses the ability of growing microtubules to push directly on the inside of the cell membrane. If the centrosome were to be much nearer one surface of the cell, even short microtubules would be able to reach all the way to that surface and push outwards against it. The centrosome would therefore experience a strong push away from this side. The opposite side of the cell could be reached only by very long microtubules, which are rare because of the ever-present danger of catastrophic failure. There would, therefore, be fewer

---

[†] 'fresh' = bound to GTP, 'stale' = bound to GDP following GTP -> GDP + Pi hydrolysis.

microtubules able to push on this side, and the centrosome would experience a much weaker push from that direction. The imbalance of forces would push the centrosome away from the nearby membrane. Only when the balance of pushing forces is equal would the centrosome settle down to a stable position, and this equality of pushing would only happen when it is equidistant from all of the surfaces; in other words, when it is in the centre of the cell (Figure 4a). Experimenters have set up centrosomes in entirely artificial boxes and have shown that they manage to find the centre of the box by this pushing mechanism.[4]

The pulling mechanism[5, 6, 7] relies on small motor proteins, known to be scattered throughout the cell, that can bind to microtubules and try to 'walk' along them towards the centrosome. By walking towards the centrosome end of a microtubule, each of these proteins exerts a tiny outward pull on the microtubules, in the way that a person walking forwards along a boat tends to push the boat backwards. The longer the microtubule, the more motor proteins can bind to it and the more pull it receives.[8] Therefore, if a centrosome is closer to one side of a cell than the other, the longer microtubules that stretch towards the far side of the cell feel more of a pull than the shorter ones, and the centrosome moves towards the centre of the cell (Figure 4b). Careful experiments in the fertilized eggs of simple animals, such as the sand dollar and the round worm, have shown that the pulling mechanism is important in these cells. When, for example, some of the microtubules are cut with a laser, the centrosomes spring back as if they had previously been held in position by microtubules under tension.[9] It is possible that both mechanisms act together in some cells, the strong pulling outwards of long microtubules reducing even further their ability to push on the centrosome, and magnifying the imbalance of pushing forces it experiences.

Whether human embryos use the push mechanism, the pull mechanism, or a combination of the two, the ultimate effect will be the same: the centrosome will centre itself automatically, without any of the components involved having to 'know' the shape of the cell and without the need for any coordinate system to specify the centre. The system will organize itself. The price of such an automatic system that can adapt itself to almost any starting conditions is the energy needed to keep building new microtubules and pulling on them; this high cost in terms of energy is typical of adaptive self-organization.

For cell division, the main theme of this chapter, the requirement is to define, not the centre of one cell, but to define the locations that will become the centres

a: the push model

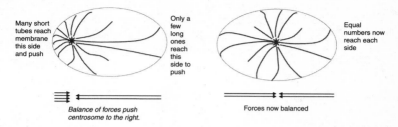

Many short tubes reach membrane this side and push

Only a few long ones reach this side to push

Equal numbers now reach each side

Balance of forces push centrosome to the right.

Forces now balanced

b: the pull model

Long tubules can recruit more motors

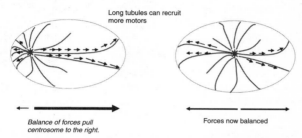

Balance of forces pull centrosome to the right.

Forces now balanced

FIGURE 4 Two theories for how the centrosome finds and uses the microtubules it emits to locate itself at the centre of a cell. In the push model (a), the microtubules push away from the cell membrane and, as there are more short tubules than long ones, the push is always strongest away from the nearest membrane of the cell. In the pull theory (b), motor proteins scattered throughout the cell pull, but many more can attach to long tubules. Since both sides of the centrosome have lots of short tubules but only ones far away from a nearby membrane can have long ones, the centrosome is pulled away from the membrane.

of the two daughter cells, so that chromosomes can be moved to them properly. Fortunately, a cell can define two 'centres' as easily as it can define one, and by the same mechanism; all it needs is to have two centrosomes.

Centrosomes consist of a 'cloud' of proteins around a pair of joined, tubulin-containing, stubbly pipe-like structures.[10] These organize the rest of the centrosomal material. In preparation for the cell dividing, the joined pair separate and, as soon as they have done so, each seeds the formation of a new partner so that there are soon two joined pairs close to one another. Each of these organizes centrosomal material around itself and nucleates the formation of new microtubules that push out into the rest of the cell, making the hub-and-spoke

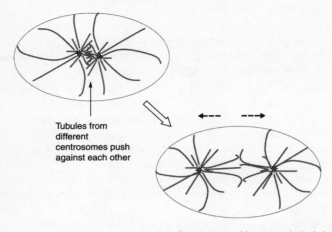

Tubules from
different
centrosomes push
against each other

So imbalance of forces results in their
separation

FIGURE 5 When there are two centrosomes in a cell, the interactions between their microtubules tend to separate them.

arrangement already described. With two centrosomes in one cell, the microtubule spoke system of one interferes with the spoke system of the other. In
terms of the pushing model, microtubules from one system will tangle with,
and therefore be able to push against, those of the other as well as against the
cell membrane. The presence of another centrosome and spoke system therefore creates a 'false impression' of how close one centrosome is to that side of
the cell, and it therefore locates not in the centre of the cell but displaced away
from the other centrosome (Figure 5). Similarly, in terms of the pull model, each
centrosome-spoke system shields the other from a pull in that direction. Both
mechanisms, which may operate simultaneously in humans, will have the same
effect: neither centrosome will locate to the centre of the cell, but they instead
take up positions about half way between the real centre of the cell and the edge
(Figure 5). In this way, the two centrosomes define the future centres of the two
new cells that will form when the original single cell divides. Again, this is automatic and neither has to 'know' anything about the detailed shape of the cell.

The microtubule spokes that radiate away from each centrosome are not just
used to define the future centres of the cells: they are also used to separate the

duplicated chromosomes so that each forming cell acquires its full complement. To achieve this, they first have to connect to chromosomes. Again, the mechanism used does not require that any components have advanced knowledge of where anything else is, and again the system makes use of the fact that microtubules are unstable structures that undergo periods of growth and then catastrophic collapse. The naked ends of microtubules are particularly vulnerable to unravelling, but if they become embedded in certain microtubule-binding proteins they are somewhat stabilized. A specific region of each chromosome contains these microtubule-binding proteins, so any growing microtubule that happens to collide, by chance, with this region of a chromosome will be protected.[11] In this way, a system in which microtubules are sent out randomly, and perish unless they are stabilized by finding a chromosome, will achieve an arrangement in which all of the chromosomes are bound, fairly stably, by microtubules.

Simple random attachment of microtubules to chromosomes may be enough to ensure that each chromosome gets moved to the centre of one of the daughter cells, but cell division demands far more than this. What is needed is for one copy of each chromosome, say the chromosome 9 from the father, to be attached to microtubules from one centrosome, while the other copy of sperm-derived chromosome 9 is attached to microtubules from the other centrosome. That way, one will go to each cell produced in the division. This arrangement is achieved by yet another influence on microtubule stability. The two copies of each chromosome, made when the DNA was replicated, are held together by special protein complexes. When these complexes are under mechanical tension, as they will be when two different sets of microtubules and their associated motor proteins are playing 'tug of war' and trying to pull the two copies to opposite centrosomes, they generate a signal that results in microtubules being stabilized very much more than if the tension is absent.[12] If both copies of the chromosome are attached to microtubules coming from the same place, this tension will be absent and the microtubules will not survive long. If they are attached to microtubules that come from different centrosomes and try to draw them apart, the stabilization signal is strong and the microtubules are much more likely to survive for a long time. The system therefore keeps changing, keeps 'exploring', until it finds the state in which sister copies of chromosomes are all being pulled in opposite directions.[13] The process is expensive in terms of energy, but is automatic and can even cope properly with extra chromosomes added by an experimenter or by evolutionary change.

Once all of the chromosomes have lined up in this way, the cell can go on to the next stage of division, in which the proteins that bind sister copies of chromosomes together let go of each other and the sisters can be moved apart towards opposite ends of the cell. This must not begin before all of the chromosomes are lined up correctly, or the daughter cells would inherit peculiar numbers of chromosomes and would be missing important genes. A system must therefore exist to prevent premature separation. This again uses the ability of proteins holding sister chromosomes together to sense the tension that is present when each site is attached to microtubules from different centres. When this pull is *absent*, these protein complexes keep producing a signal, a particular type of small molecule that can spread through the cell, that blocks progression to the next stage of cell division. Essentially, they are yelling '*not yet!*' in the language of biochemistry. As long as any chromosome has still to be properly connected and pulled on, this '*not yet!*' signal will be made and the cell will wait. Only when all are correctly tensioned will all of the signalling complexes have fallen silent, and the next stage will be ready to begin. Again, this system can adapt itself to accommodate any number of chromosomes.

When all of the chromosomes are properly lined up in the middle of the spindle, and all of the '*not yet!*' signals have faded away, the cell is able to proceed to the next stage. The protein complexes that hold each pair of sister copies together let go of each other, leaving the motor proteins free to pull the chromosomes up the microtubules towards each centrosome.[14] Once all of the chromosomes have moved, further automatic systems place a ring of contractile proteins around the 'equator' of the cell between the poles defined by the centrosomes. These proteins slide across each other, and this folds a 'waist' into the cell. The waist continues to contract until the cell if finally split completely into two new cells.

Seen as a whole, the systems described above may seem very elaborate and complicated. Seen component-by-component, though, they are very simple. Each component protein has only to do a very simple task. The ability of the system as a whole to do complicated tasks, such as finding and separating chromosomes accurately wherever they might initially lie, emerges from the connections between simple components, rather than from any complications in the components themselves. In particular, it emerges from the way that information about what the system has achieved, for example whether all of the chromosomes are lined up yet, is fed back to the behaviour of its components. The use of simple component parts connected together with rich feedback is characteristic of life, and this chapter has gone into details precisely to convey a feeling of how systems of 'dumb' biological molecules can be organized to solve problems: a feeling of how the simple can give rise to the complex.

The systems used to drive the first cell division are used again and again in the embryo, and from now on we can take them for granted. Again this is typical of biology—when something works it tends to be re-used, sometimes adapted a little, as embryonic development proceeds. Almost as soon as the first cell division is completed, each of the two cells begins to copy its chromosomes and to divide so that there are four cells in the embryo. This pattern continues for a while, although different cells tend to drift with respect to the exact timing of division, so from about the sixteen-celled stage the perfect multiples-of-two progression becomes blurred. Most of the time, all of the cells produced by cleavage of the early embryo remain stuck loosely together. Occasionally, in around one in twelve hundred births, the cells fall apart into two separate clumps. Each clump will go on to make a complete embryo of its own, and these embryos will each have its own placenta and be surrounded by its own membranes. This is one of three ways in which identical twins can arise, and it accounts for about a third of cases. The fact that an early embryo can split like this and produce two babies tells us something very significant about early development: all of the cells must be equally capable of making any part of the body, and there is no special cell that is either in charge or committed to making one part (the head, say). If cells were already different, if one or more of them were already committed to making some part, or if one cell were in charge, then splitting the clump would leave at least one of the resulting groups of cells without a committed cell type, or without the cell in charge, and its development would fail. Each one of the thousands of identical twins in the country is an eloquent testament to the equality of opportunity enjoyed by cleavage-stage cells.

Once the embryo has reached about the sixteen-cell stage, it has enough cells to organize a new shape for itself, and to make its cells start to become different from one another. At this point, the prelude is over, and the real work of embryonic development begins.

3

MAKING A DIFFERENCE

Honest differences are a healthy sign of progress.
Mohandas Karamchand Gandhi

The main achievement of the preceding cleavage stage of development described in Chapter 2 can be captured in the phrase 'more of the same'. Cell number increases exponentially, but every cell produced is identical to every other. Indeed, for the first few divisions none of the cells even bothers to make use of its own genes, relying instead on stores of molecules laid down in the egg by the mother and shared out amongst the cells produced when the egg divides.[1] Dedicating early development to cell multiplication with no other distractions makes sense: the more cells that exist, the easier it will be to make a body with them when the time comes. The game can be played only so far, though, because cleavage means that each generation consists of cells that each have half the volume of their immediate predecessors; soon the size of cells will reach a practical minimum and interludes of cell growth will be needed between divisions. Cell growth implies a need for nutrients and the requirement to do something about obtaining them. This, in turn, means that some cells need to specialize to bring food to the others. In many animal embryos, feeding is achieved by harvesting yolk laid down in an egg. In mammals, it involves bringing resources directly from the mother, but the general point holds true: at some point, cleavage must end and specialization must begin.

Cell specialization implies that a collection of initially identical cells will lose its homogeneity so that some cells do one thing and others do another. This brings the embryo up against a fundamental problem. Creating a difference

28

means creating new order and new information. The increase in information content is reflected in the fact that it takes more words, or more mathematical symbols, to describe the shape of an asymmetrical object than a symmetrical one. Describing the shape of a cup with a handle is, for example, more involved than describing the shape of a cup without one. In many 'lower' animals, such as insects, the mother provides the necessary information. She copies spatial information from her own body into the egg, in the form of concentration gradients of specific molecules, so that the cells of the embryo inherit different quantities of these molecules as the egg divides. The cells are then able to use these differences to decide to take different paths of development. This non-genetic method of passing critical information down through the generations is highly effective but, as far as we know from many fruitless attempts to find it in mammals, humans do not rely on it and one side of the human egg is much like the other.[2] The creation of a difference—which means new information—therefore poses a serious logical problem to the human embryo: how can a pattern be created where there was none before? The embryo's solution to the problem is elegant: it draws information from the laws of geometry.

When there were only a few cells in the embryo, every one took up a large enough portion of the total volume that some of each cell's surface faced the outside. Once cleavage has generated thirty-two or sixty-four cells, though, they are small enough in relation to the size of the spherical embryo that some find themselves in the middle, completely surrounded. Others are still on the outside, with around a sixth of their membrane facing clear space. Cells can sense whether they are completely surrounded by other cells or whether parts of them face nothing more than fluid, and they use this information to decide what to do next. Those that find themselves with a free surface activate a set of previously inactive genes and become the embryo's first specialized tissue, the trophectoderm. Cells with no free surface, on the other hand, hold those genes firmly 'off'. This use of a simple physical cue (free surface) as something that can be interpreted as 'information' frees the embryo from any requirement to have a prior spatial plan. It also liberates cells from any need to know exactly where they are in the embryo: all they need to detect is whether they have a free region of their surface or not.

The trophectoderm is concerned only with building structures to support and feed the rest of the embryo: it will never become part of the baby itself.[3] The first thing the trophectoderm does is to pump fluid into the embryo. This fluid accumulates to form a large, watery cavity (Figure 6). The formation of this

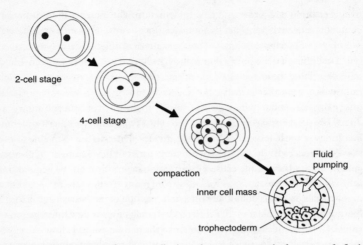

FIGURE 6 The progression from two cells, through compaction, to the formation of a fluid-filled cavity with trophectoderm and inner cell mass becoming different from one another. The 'capsule' drawn around the whole embryo is a tough jelly-like coat called the zone pellucida that was originally the coat around the egg.

cavity leaves the inner cells as an off-centre clump, the inner cell mass, stuck to the inside of one part of the tophectoderm. Judged by their appearance alone, this seems much less interesting than the active, pumping, invading, foraging cells of the trophectoderm, but it is the inner cell mass that will go on to form the baby. Sometimes, the inner cell mass does not form as a single entity but rather as two separate clumps, each of which will go on to form its own individual human. Like the twins mentioned in Chapter 2, these twins will be genetically identical, but this time they will grow up in the same shared trophectoderm, although they will each make their own yolk sacs and amniotic cavities (see later), and so will remain safely separated. This is the most common way that identical twins arise, being responsible for around two-thirds of cases (the third way, described later in this book, is very rare).

By the time that its outer cells are transporting fluid and making an inflated structure, the embryo should have left its mother's fallopian tube, where fertilization took place, and entered her womb. This organ is normally a fairly 'deflated' structure, with its walls close together in the manner of a rubber glove with no hand in it; this is especially true in the week after ovulation. The young embryo is therefore very likely to encounter its inner surface soon after it enters. When it does, it uses a special set of adhesion molecules to attach itself. Once the cells have stuck, they make new proteins that allow them to push in between the womb's own cells.[4] Within a few hours, 'fingers' of lined-up embryonic cells will have pushed, like an advancing army, into the mother's tissue to make the placenta. Many of the womb's cells are destroyed in the process, and their remains are taken up by the embryo as much-needed food. The response of the mother to this assault causes the death of even more of her tissue so that, by ten days after fertilization, the destruction has formed an ulcer-like cavity so large that the whole embryo lies within it. In humans and in some other animals, the inner lining of the womb grows back over it, healing over the hole.

The mode of life of a human embryo is essentially parasitic, but this metaphor must not be taken too far. The embryo may be a parasite but, since tolerance of this parasitism is the only way that a woman can reproduce, it is essential to the survival of the species. Indeed, the mother does not merely tolerate parasitism by her offspring but encourages it: if the dialogue between the signalling molecules made by a mother's womb and the signals from the embryo is interrupted, the embryo can no longer implant itself and pregnancy fails.[5]

Sometimes, the movement of the embryo along the female reproductive tract, from the site of its fertilization in the oviduct (Fallopian tube), is abnormally slow and it does not reach the womb by the time it is ready to implant. One common cause for this slow transit is damage to the oviduct lining by *Chlamydia*,[6] now very common in young adults.* The growing embryo attempts to implant wherever it happens to be when it reaches that stage of development and, if it finds itself still in the oviduct, it will try to establish itself there. This produces an ectopic pregnancy. The oviduct is not adapted to support the growth of a foetus, either physically in terms of size and ability to stretch, or physiologically in terms of its ability to provide food and blood. In many cases, the pregnancy fails, and this is a common reason for miscarriage. In other cases it may be necessary to induce miscarriage to protect the mother's life.

Before they begin their own programmes of specialization on the way to making the baby itself, the cells of the inner cell mass have the potential to form any cell of the body (at least in mice: the experiment cannot be done in humans

---

* Chlamydia is not the only cause of ectopic pregnancy, of course: many cases remain completely unexplained.

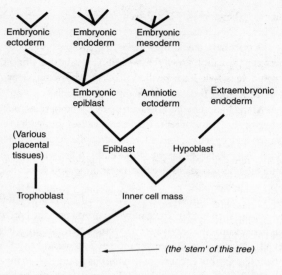

FIGURE 7 A typical tree-diagram that summarizes the routes by which cells of an early stage of development give rise to (turn into) cells of more mature stages. The diagram is read upwards. This particular diagram concerns the events in this chapter and the next. A complete tree for all human development would extend upwards a long way and have hundreds of branches. For each diagram like this, whether it starts off at the beginning of development or begins with one of the cells that is already some way up the complete tree, the lowest point of the diagram is considered the stem and the cells there are 'stem cells' for the part of the tree depicted. The main text explains more about the correct use of the term 'stem cell'.

for ethical reasons). When researchers want to describe the range of cells that a cell can make, they often depict these cells, and the order in which they appear, in a diagram that looks like a branched tree. Figure 7 shows a small example. In such a diagram, all possibilities radiate, ultimately, from one cell type that is the 'stem' of the branched tree. These cells are therefore referred to as 'stem cells' (a translation of *Stammzelle*, the name coined in 1909 by Alexander Maximow to describe cells at the base of the 'tree' of blood cell types). Depending on the diagram in question, the term 'stem cell' may be applied to a cell that can make only a few different types of cells of the body or to cells that can make a larger range of types.

Recently, use of the word 'stem cell' has acquired a slight additional restriction, in that many authors insist it be applied only to cell types that can maintain their own population as well as give rise to the cell types above them on the tree (this was not an explicit feature of Maximow's original use of the term[7]). Fortunately, the subtle re-definition of 'stem cell' makes no difference to our applying the term to the inner cell mass, near the bottom right of Figure 7, because inner cell mass cells can renew their own populations even when removed from an embryo and kept in a culture flask. They will give rise to all of the tissues of the body. They are therefore sometimes referred to as embryonic stem cells, or ES cells, especially when they have been taken from an embryo and are growing in culture.[8]

Embryonic stem cells have made a major impact on biomedical research over the past decade or so. Scientists routinely make specific, designed alterations to the genes of mouse ES cells and then inject some of these cells into the inner cell mass of a normal mouse embryo. The mouse that results has a body that is a mixture of normal cells, from the un-manipulated inner cell mass, and the genetically modified cells from the engineered ES cells. In general, the mixture is fine enough that some of the sperm of a male mouse will develop from the genetically modified cells. The offspring of that mouse, made by normal mating, will include animals that were made by one of those sperm and therefore carry the modified genes in all cells of their body. This technique has allowed scientists to make animal 'copies' of human genetic diseases, so that they can perform experiments aimed at a better understanding of the disease and how it might be treated.[9] Tens of thousands of such mice have now been made, and many of the facts quoted in this book were established by observing the effects on pregnancy and embryonic development of mutating or deleting specific genes.

Human ES cells also exist.[10] There is hope that, if we find ways of controlling their development to give rise to specific cell types, they may be used to repair damaged human tissues or to make new ones for transplant. That said, the existence of such cells is not universally welcomed because their founder colonies can be obtained only through the destruction of an early human embryo. For some people, the fact that an early human embryo has the potential to become a human being means that it should always be accorded full human status. To those who hold these views, destruction of a human embryo is unacceptable, effectively being murder, no matter what good may come from it. For others, the early embryo is so far from having the attributes of a human being, such as thought or feeling, that it deserves no special protection. Yet others take a position somewhere between, and allow human embryos to be used for research as long as their use is justified according to an agreed code of principles. A recent development may provide a way out of this ethical dilemma. A research team in Japan recently discovered a way to turn ordinary body cells of a mouse into ES-like cells, which they call iPS cells.[11] To prove how like ES cells these are, the researchers used them to replace the normal inner cell mass of an embryo and succeeded in making mice from them. The technique is relatively easy and is now used in many labs around the world. Many people have applied it to adult human cells and have made what appear to be human iPS cells from them ('appear to be' because the ultimate test of building a complete human from them cannot be done, as it could be for the mouse, for ethical reasons). If iPS cells really are identical to human ES cells, the argument goes, they could be used for anything human ES cells would be used for, and thus the same medical advances would be made with no need to destroy human embryos. This may be true, but it may not necessarily be good news for those who accord full human status to the embryo on the basis that the embryo has the potential to become human. If all of our cells have the potential to be iPS cells and to become human, what is to stop full human status being applied to them too?

In a normal embryo, cells of the inner cell mass cells are in that state for only a very short time before they begin to specialize and become different from each other. The presence of the fluid-filled cavity in the embryo means that the inner cell mass, which does not increase in volume much, has some cells that are completely surrounded by others but some that now have free surfaces. Again, the free surface breaks the symmetry of cell–cell adhesion and makes the layer facing the fluid distinct. This becomes a tightly sealed sheet, called the hypoblast.† It used to be thought that cells chose to become hypoblast as a direct consequence of finding themselves at that location, but recent experiments in mice have suggested that some cells of the inner cell mass might be randomly pre-disposed to make hypoblast. Once the free surface is available, these cells migrate towards it and settle there, the others retreating inwards.[12] Either way, it is again a free surface that is used to impose spatial order (Figure 8). Some hypoblast cells remain where they form but many spread out to line the trophectoderm layer and form a hollow bag called the yolk sac, a name that is used for consistency with the embryos of 'lower' vertebrates (Figure 8).

---

† The hypoblast is commonly called the 'primitive endoderm' where mammals are involved. This book uses the word 'hypoblast' to avoid possible confusion with another type of 'endoderm'.

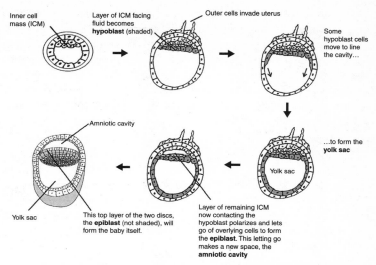

Inner cell mass (ICM)

Layer of ICM facing fluid becomes hypoblast (shaded)

Outer cells invade uterus

Some hypoblast cells move to line the cavity...

...to form the **yolk sac**

Amniotic cavity

Yolk sac

Yolk sac

This top layer of the two discs, the **epiblast** (not shaded), will form the baby itself.

Layer of remaining ICM now contacting the hypoblast polarizes and lets go of overlying cells to form the **epiblast**. This letting go makes a new space, the **amniotic cavity**

FIGURE 8 The embryo uses the free surface trick again to make the layer of the inner cell mass that faces fluid make a new cell type, the hypoblast. The layer of the remaining inner cell mass that now finds itself lying against the hypoblast becomes different again, forming the epiblast. This lets go of overlying cells, so that a new cavity is made. The epiblast, an unassuming disc of cells, will give rise to all of the baby; everything else makes tissues that support life in the uterus.

The remaining parts of the inner cell mass cells then separate into two layers. The cells immediately adjacent to the hypoblast stick to it and remain as a new layer, called the epiblast. The cells that overlie this pull away, making a new cavity where they part company from the epiblast. This is called the amniotic cavity (Figure 8). The double-layered epiblast–hypoblast disc is therefore left sandwiched between the yolk sac cavity and the amniotic cavity, like the cross-line in the Greek letter θ. The layer facing the amniotic cavity, the epiblast, will be the source of all of the cells in the baby.[13, 14]

The events of this chapter were concerned mostly with the breaking of sameness and with the establishment of differences between previously identical cells. The same trick, that of using a free surface, was used again and again in this early phase of development because it afforded a way of using geometry to provide new information. In each case, purely local influences caused larger-scale changes without any cell having to be in charge or to perceive the big picture. Once different tissues already exist, it becomes easier for an embryo to create further differences for itself. For example, a new cell type, C, might develop where existing cell types A and B are in contact, and this would create two new zones of contact (A-C and C-B), each of which could then be used to specify yet more types of cell. The era of using mainly 'external' influences, such as free surfaces, is therefore replaced at later stages by mechanisms that use internal differences instead. The first of these, arguably the most remarkable of all, will be the subject of the next chapter.

4

LAYING DOWN
A BODY PLAN

It is not birth, marriage, or death, but gastrulation which is
truly the most important time in your life.

Lewis Wolpert

When people of mature years look back on the lives they have led so far, most perceive long periods of routine living punctuated by short bursts of rapid change. The changes may well be the result of gradual preparation over many months or years but, in our perception, the preparation is hard to see. A baby's babbling becomes more complex so gradually that his parents barely perceive the change, yet they never forget his uttering a complete word. A couple's formation of a lasting bond is a hidden process, a slow accumulation of trust and shared vision that most people notice only in retrospect when they have realized what they have come to mean to one another. Professional skills accumulate steadily in a process that is much less obvious than the new job or promotion that is their reward. Less happily, creeping accumulation of cellular damage is scarcely perceptible until it crosses a threshold to become a definite, diagnosed disease, and its subject crosses that subtle line from well person to patient.

This alternation of apparent constancy with bursts of change is as much a feature of an embryo's life as of an adult's. In the embryo's very early life, covered in the last two chapters, there was first a period of simple cell division with nothing much else going on, then an apparently sudden switch to the new

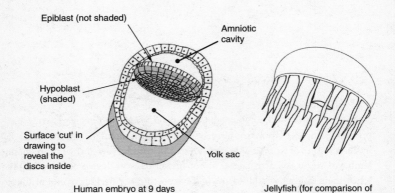

Epiblast (not shaded)

Amniotic
cavity

Hypoblast
(shaded)

Surface 'cut' in
drawing to
reveal the
discs inside

Yolk sac

Human embryo at 9 days

Jellyfish (for comparison of
symmetry)

FIGURE 9 The simple, radial symmetry of the epiblast-hypoblast discs in a human embryo, compared with the similar symmetry of a jellyfish.

activity of making layers of cells different from one another. The structure that results from this, essentially a couple of discs crossing a fluid-filled sphere (Figure 9), is still a very long way from being recognizable as anything remotely human. If a reasonable person were told that the pair of discs were to produce any animal, he would probably assume it must be a jellyfish of some kind, as the bell of a jellyfish does at least have the same radial symmetry as a disc, with a top and a bottom but no special axis running across from edge to edge (Figure 9). The embryo has, though, laid down all of the preparations necessary for a sudden and massive re-organization from which, in a matter of two days, it will become something recognizable as a proper body. The process is called gastrulation.

One way of approaching the topic of gastrulation would be to consider first the structure of the primitive body that will result from the process, and then to illustrate how each event in gastrulation contributes to making that body. This way of writing would perhaps be the easiest to follow but it would convey a false implication that the cells themselves have some idea of the anatomy of the body they must build, and that they work towards it. Real development relies not on individual cells having a knowledge that even we, with our brains of hundreds of millions of cells, find difficult to grasp, but rather on their responding in a simple, automatic way to the environment that surrounds them. The chapter will therefore follow development forwards, with the body emerging gradually from the behaviour of the cells, and only at the end will we take stock of what has been achieved.

Before the story, an important warning: the process of gastrulation is extremely difficult to study in humans so almost everything in this chapter has had to be inferred from what happens in related animals. There are strong, legal restrictions on how far scientists are allowed to grow human embryos outside the body, and the stage of gastrulation lies beyond the legal limits, for reasons that will be explored later in this chapter. The basic sequence of anatomical changes in humans is known, having have been worked out from relatively few, precious samples of human embryos that have been found in post-mortems or hysterectomies of women who generally did not even know they were pregnant (embryos begin to gastrulate at about fifteen days of development, roughly when their mother would normally be expecting her period). Some of these samples are now well over a century old, but are still looked after with great care in museums because they would be so hard to replace. The topic receives intense attention in mice and chickens, neither of which have anatomies exactly like humans at gastrulation: chickens grow in eggs rather than a womb, and mice have epiblasts and hypoblasts that are cup shaped, rather than disc shaped, which may make a difference. Mapping the mechanisms that have been discovered in these experimental animals to the human therefore involves a greater risk than usual of making an assumption too far and getting the details wrong.

The starting point for gastrulation is the embryo as we left it at the end of the Chapter 3. By that stage, it had formed a number of supportive tissues, such as the placenta, and had formed two fluid-filled internal spaces, the amniotic cavity and the yolk sac. Between these cavities were two discs, lying one above the other; the hypoblast, which will give rise to yet more supportive tissues, and the epiblast, which will give rise to the baby. Neither has any overt distinctions that make one part of the edge of the disc different from any other part (Figure 9).

The first change that has so far been detected (in experimental animals) takes place in the hypoblast. Here cells in the middle of the disc* begin to switch on new genes, including one that specifies the production of a DNA-binding protein called Hex.[1] The triggers for this change, and for its location, are not yet known: one possibility is that hypoblast cells are primed to make the

---

* This translation from mouse (distal pole) to human (centre of hypoblast) anatomies is a best guess, and may turn out to be wrong if proper human data are one day obtained.

change anyway, but that most of them are inhibited by a signalling protein that is made by the supportive tissues that surround the hypoblast edge.[2] Only in the centre of the disc would cells be far enough from the source of this inhibitory protein to be able to escape its effects, and be free to switch on Hex. This mechanism is conjecture, but the switching on of Hex is not. The cells that make Hex move away from their point of origin, by pushing their way past their neighbours, and congregate at one point on the outside of the hypoblast disc[3, 4, 5] (Figure 10). It is still not at all clear, even in mice, what is so special about this point in the disc that the moving cells choose to congregate there. In 'lower' animals, the equivalent location is specified by prior events. In some cases, it is marked by an asymmetrically located cue left in the egg by the mother, and kept as an asymmetry in the embryo. In others, it is marked by the locations of polar bodies, which are 'waste products' of the cell divisions that make the egg. In yet others, it seems to be marked by the place that the sperm arrived. It is possible that a system like this operates in mammals, and there is some evidence of asymmetry of even very early embryos, at only the stage of cell compaction, in mice.[6] Unsurprisingly, nothing has yet been shown convincingly for humans. Our lack of knowledge on this point is frustrating, because the accumulation of Hex-expressing cells at one place achieves something extremely important. It specifies one place around the rim of the hypoblast disc as 'different': it is therefore the first visible break in the simple, radial symmetry of the embryo (Figure 10).

Once they have reached the rim of the hypoblast, the collection of Hex-expressing cells becomes known as the AVE.[†] Cells of the AVE start to secrete their own signalling proteins. These proteins spread over short distances and can therefore reach the epiblast that lies immediately above the hypoblast, well within signalling range.[7] By this stage, cells of the epiblast are already responding to signals coming from the support structures of the embryo, and these signals are causing them to prepare to make structures characteristic of the posterior end of the body. If nothing else were to happen, the whole epiblast would begin to make posterior structures, which would be a disaster for the embryo. The signals coming from the AVE, however, oppose the tendency to make posterior structures and induce the epiblast cells to activate genes that will lead them to make head tissues instead.[8] If the AVE fails to make its

[†] The expansions of all abbreviations are given at the end of this book, but the words for which both 'V' and 'E' stand in this particular term are much more likely to cause confusion than to help, which is why this chapter will use the abbreviation only.

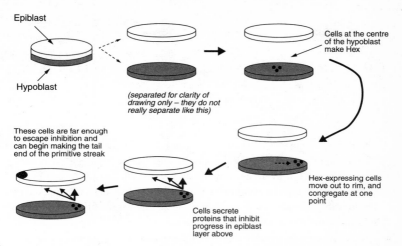

Epiblast

Hypoblast

(separated for clarity of drawing only – they do not really separate like this)

Cells at the centre of the hypoblast make Hex

Hex-expressing cells move out to rim, and congregate at one point

These cells are far enough to escape inhibition and can begin making the tail end of the primitive streak

Cells secrete proteins that inhibit progress in epiblast layer above

FIGURE 10 The breaking of radial symmetry. A group of hypoblast cells at the centre of the hypoblast switches on Hex and moves to the edge, to congregate in one place. This breaks the radial symmetry of the hypoblast disc and, because the cells secrete proteins that interfere with signalling events in the epiblast above, it breaks the symmetry there too.

signalling proteins, the embryo fails to form a proper head. The position of the AVE, at one unique place around the circumference of the disc, therefore sets up a polarity in the overlying epiblast, the side nearest the AVE becoming the most head-like and the part furthest from it becoming the most posterior. Only the part of the epiblast disc furthest from the AVE is distant enough to escape its signals, and this seems to be important in fixing a location for the first directly visible change in the epiblast. In this location, epiblast cells switch on the production of a signalling protein that attracts surrounding cells to move towards them.[9] The result of this attraction and movement is the beginning of a very important structure with an unassuming name: it is called the primitive streak (Figure 10).

The importance of the initial action of the AVE cells down in the hypoblast to setting out where the primitive streak will form has been demonstrated in chick embryos by allowing symmetry-breaking to begin and then rotating the hypoblast to a new position.[10] When the hypoblast is rotated, the epiblast arranges the formation of its primitive streak to match, confirming the hypoblast is effectively in control of direction.

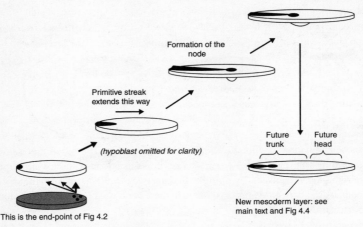

Formation of the node

Primitive streak extends this way

(hypoblast omitted for clarity)

Future trunk   Future head

New mesoderm layer: see main text and Fig 4.4

This is the end-point of Fig 4.2

FIGURE 11 Elongation of the primitive streak and formation of the node (a term that will be explained later in this chapter).

As more epiblast cells converge, they pile up and jostle each other, gradually elongating the primitive streak inwards along the radius of the epiblast disc (Figure 11). As more cells join in, they reduce the population that remains at the sides of the disc so, as the streak lengthens, the disc as a whole begins to narrow. What was a round disc therefore becomes an oval, and the simple radial symmetry of the young epiblast is lost (Figure 11). The long axis of the oval is the first sign of the long axis of the body, the axis that runs from the top of the head to the far end of the spine (in most animals, it would be natural to say 'to the tip of the tail', but that expression always sounds a little odd for humans). The first part of the primitive streak to form, the part nearest the outside of the disc, defines the posterior. The last part of the streak to form, near the middle of the now-oval disc is near what will be the head. Discovering that we make the first visible sign of our body where our bottoms will be, and that we leave sculpting a head until later, is one of the many ways in which embryology reminds us humans not to take our dignity too seriously.

The role played by the AVE in specifying where a primitive streak will form in the epiblast above has an interesting consequence. If, by rare chance, the Hex-expressing cells fail to congregate in one place on the rim but rather make two separate small AVEs, there will be two separate signalling centres. Under such circumstances, it is possible for two separate primitive streaks to form. This produces the third, and rarest, type of twinning, seen in fewer than 1 per cent of identical twin pregnancies.

Where two complete, independent primitive streaks form, they will make two completely separate head–tail axes and therefore two bodies. These twins, arising from the same epiblast long after the amniotic cavity formed, will share the same amniotic cavity as well as sharing the same chorionic cavity. This makes them unlike the twins mentioned in Chapter 2, who shared neither, or those in Chapter 3, who shared a chorionic cavity but not an amniotic one (counting how many cavities twins share is how obstetricians know which type of twinning is involved). Twinning by making two primitive streaks in the same epiblast is a dangerous process, because there is no clear line of demarcation between the two bodies and there is a serious risk that they will never fully separate. If separation fails, the twins will therefore be born conjoined—connected and sharing at least some portions of their bodies. Often, these can be really important parts, such as major internal organs. Depending on the anatomy that results, conjoined people can go on to live healthy, if somewhat complicated, lives. Chang and Eng Bunker[11, 12] (1811–1874) are probably the most famous conjoined people ever to have lived; they toured for many years with P. T. Barnum's circus during the nineteenth century, when the age of the freak show had not yet passed and most circuses would feature bearded ladies, dwarfs, giants, and morbidly obese people to be gawped at by a curious, unsympathetic public. The Bunker brothers named their act after their country of origin, and gave the world its popular name for conjoined people: 'Siamese twins'. Were they to be born today, the Bunker brothers could probably be separated, but many other instances of conjoining are more complicated and separation may be impossible, or may be possible only at the expense of one twin's life. Such cases pose severe ethical, as well as surgical, challenges.

Even more complicated situations arise when the epiblast forms not two separate primitive streaks, which give two distinct body axes even if conjoining has taken place, but forms a single 'Y'-shaped, incompletely duplicated, primitive streak. This results in two heads, and perhaps necks, that share a common trunk. Confusingly, this phenomenon is usually called an 'axis duplication' (confusingly because the whole point is that the axis does *not* completely duplicate, which would be fine, but duplicates only partially to result in the Y-shape). Human examples are rare and usually aborted, stillborn, or short lived; most long-standing anatomical and surgical museums have examples floating about in Victorian jars. There are some rare survivors, most notably Abigail and Brittany Hensel, young women now in their early 20s, each of whom has a head and neck of her own but who share a body. Each can perform 'head' functions, such as reading, entirely independently. They cooperate for tasks such as walking, playing the piano and driving (devising an appropriate individual driving test posed an interesting challenge for state bureaucracy). Unusual anatomy aside, the Hensels are entirely normal, intelligent, sociable young women. Axis duplications are more common in other animal types, such as reptiles and amphibians, and show a better survival: a two-headed rat snake named *We* lived for eight years in St Louis City Museum and became one of its star attractions during the early years of this century, both for its peculiar anatomy and for the occasional arguments that would go on between its heads, each of which had an independent will. One example of an adult two-headed lizard has even turned up in fossil form.[13] Axis duplications can be induced to occur in amphibians by deliberate interference with the signalling mechanisms that are the frog equivalents of signals within the epiblast–hypoblast system of a mammal; indeed, this is one of the ways in which the signalling mechanisms were identified in the first place. It was in frogs, for example, that one of the proteins made by the mammalian AVE, and necessary for head formation, was shown to be able to induce extra heads when injected into frog embryos: the protein was therefore called Cerberus, after the multi-headed canine guardian of the gates of Hades.

The association between one primitive streak and one individual person has been interpreted by some bio-ethicists as indicating that formation of a primitive streak is a critical ethical boundary in human development. The argument goes that, before a primitive streak has formed, it is not certain how many individuals will come from an embryo so one embryo cannot be equated with one human being. By extension, the argument goes on to say that any entity that cannot be equated with an individual human being cannot be given human rights and, by further extension, that it is acceptable to manipulate embryos up to this stage but not beyond. Conversely, once at least one primitive streak has formed, there is a definite number of identifiable individuals, so these have individual human rights and cannot therefore be experimented upon. Lines of reasoning like this, or like any of the others that set boundaries at other stages, including conception itself, suffer from the basic problem that law-makers want to see a clear boundary between not-yet-a-person and now-a-person. Some aspects of development really do involve step-changes, as described at the beginning of this chapter, but others (size, for example) show only gradual change. It is possible that person-hood is not a step-change, but that the potential for humanity is converted to actual human nature over a sequence of steps that take at least months and, arguably, years (the wiring of the brain continues long after birth). The truth is that we do not yet understand enough about the biological basis of our humanity, our person-hood, to be able to ask whether it arrives suddenly or grows gradually. That is why current ethical arguments designed to draw a sharp line in the sands of developmental time verge on sophistry.

The movements that bring cells together to form the primitive streak do not stop with its formation. As more cells come in, the centre of the streak dips down to form a narrow valley, and the end that is near the middle of the disc makes a larger flat depression, called the node. Cells of the node make a number of signalling proteins that both attract further cells, and cause them to weaken their cell–cell adhesion systems and switch on genes that will drive them towards becoming other cell types. The cells closest to the source of signal receive the largest dose so respond most strongly. The loosened cell–cell adhesion and enhanced migration frees responding cells to move and to leave the sheet-like disc that surrounds them.[14, 15, 16] They do this by dropping under the epiblast disc. The process of dropping spreads tailwards along the streak, so that more posterior regions of the body undergo it considerably later than the upper trunk. As the nearest cells drop through, their old neighbours move in to close up the space, and these then dive through in their turn.

The first cells that drop through the primitive streak at a particular level of the head–tail axis join the hypoblast layer and push the existing hypoblast cells aside, leaving the central axis of this layer composed of the new cells (Figure 12). Once in position, this sheet of cells is called the endoderm (= 'inner skin', so called because it will go on to form the tube of the gut and the tubes of its associated organs such as the liver and pancreas). Cells that drop through later remain only loosely associated with one another and do not form a sheet; instead, they make loose packing material called mesoderm (meso = 'middle', because this layer is between the endoderm and the final layer). Cells that never manage to drop through and remain in the top layer after the node has passed become ectoderm (= 'outer skin').[17] The primitive streak and node therefore not only demonstrate a head–tail axis in the embryo, they also turn what was one layer of tissue, the epiblast, into three distinct layers, the ectoderm, mesoderm, and

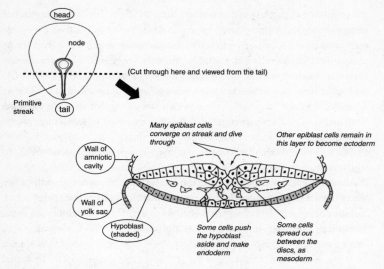

FIGURE 12 Formation of the three body layers, ectoderm, mesoderm, and endoderm during gastrulation. The sketch at the top left of the diagram depicts the epiblast of an embryo that has already formed its node, looking down from the amniotic cavity. The hypoblast is hidden underneath. The main drawing depicts a section through the primitive streak, along the dotted line. Cells of the epiblast are converging on the valley-like primitive streak and diving down through it to become part of the endoderm, which replaces the hypoblast cells and pushes them out to the edge, or the mesoderm, which makes a new layer in the middle. The cells that remain in the top layer become ectoderm.

endoderm (Figure 12). These are the fundamental layers of the bodies of almost all animals.[‡]

Because cells have to move to the primitive streak in order to dive down, the time at which they dive is related closely to their initial distance from the streak. Cells close to the mid-line have only a short way to go and dive down early, whereas cells further to the sides can dive only later, when the nearer ones have already gone. The fact that time and space are linked so intimately in gastrulation makes it difficult to disentangle which is the more important in specifying what a cell will become. Is fate already determined, before cell movement has even started, by the first signals from AVE, or is it set only as cells make their dive for the inside of the body? In some cases, it seems clear that cells are

---

[‡] A few primitive types, such as jellyfish, have only ectoderm and endoderm.

pre-programmed with their fates before they move:[18, 19, 20, 21, 22] the programming of head fate by signals from the AVE, which has already been mentioned, is an example, as is the pre-programming of cells very distant from the AVE to make primitive streak. It is not yet clear, though, how finely the future careers of other cells are defined at this very early stage. What complicates analysis further is that some 'pre-programming' is really just a bias in what cells are likely to do, and if they are deliberately placed in a different environment, to simulate what might happen if they somehow got lost or swept up in the wrong stream of cells at gastrulation, they can 'change their minds' in response to the new signals around them. The whole area is something about which embryologists still have much to learn.

Almost as soon as the endoderm layer has formed, cells from its centre-line, below the site of the primitive streak, remove themselves from their neighbours and move upwards (Figure 13). These are the cells that dived through the head side of the node itself and therefore experienced the highest concentrations of

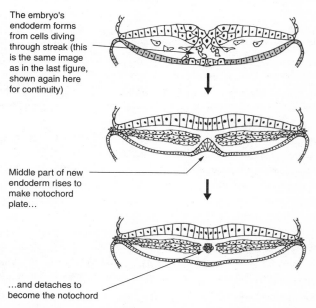

The embryo's endoderm forms from cells diving through streak (this is the same image as in the last figure, shown again here for continuity)

Middle part of new endoderm rises to make notochord plate...

...and detaches to become the notochord

FIGURE 13 Formation of the notochord from cells on the centre-line of the new endoderm that was created by cells diving through the primitive streak.

the signalling molecules that its cells produce. These signalling molecules 'pre-programme' the cells to prepare for leaving the endoderm layer again.[23, 24, 25] Once they have left, they line up along the embryo and form a stiff rod, called the notochord.§ The notochord is one of the most important structures we make as early embryos. The reason we bother to make it so early, indeed the reason we bother to make it at all, may be understood most easily in the light of evolution.

Zoologists classify the animal kingdom according to a hierarchical scheme. The fundamental unit is the species (e.g. Homo _sapiens_); similar species are grouped into a genus (e.g. Homo); similar genera are grouped into a family (e.g. Hominidae); similar families are grouped into orders (e.g. Primates); similar orders are grouped into classes (e.g. Mammalia); similar classes are grouped into a sub-Phylum (e.g. Vertebrata); and sub-phyla are grouped into a Phylum. The classification scheme was based originally on similarity, with no idea of common descent, but since Darwin and Wallace provided a theory of descent with modification and natural selection to explain why the patterns of clustered similarity exist, the schemes are usually taken to indicate evolutionary relationships as well. The phylum that includes the vertebrates is Chordata, the set of all animals that make a notochord at some time in their lives. In terms of numbers, the vertebrates dominate Chordata, but a few invertebrate types linger on from what may have been a much more impressive range back in the early Cambrian era. Most of these are rather obscure organisms, but the lancelet is common enough to be used for food in some parts of Asia. Lancelets are approximately fish-shaped animals about two inches long. They have no bony skeleton but they retain their notochord throughout adult life to stiffen the body and provide something against which muscles can work. It is therefore essential to them, from the swimming tadpole stage onwards, and it is important for the tadpole stages of other invertebrate chordates.

The notochord is, however, more than just a source of stiffness. Being composed of a distinct type of cell, which secretes distinct proteins, the notochord has the potential to be used as a source of signals to organize the embryo. Indeed, the fact that it runs straight down the centre-line makes it ideal for that purpose. Chordates, including vertebrates, make extensive use of these signals to organize their internal tissues, for example to determine the different types

§ In most animals, the notochord comes directly from mesoderm. The discovery that it comes from the middle of the endoderm layer in mice, and presumably humans, was a fairly recent surprise.

of nerve cell that form in the spinal cord, or the different types of connective tissue and muscle that form down both sides of the body. The importance of notochord signalling is underlined by the fact that it will play a major role in the events described in three later chapters of this book (5, 7, and 9). Vertebrate evolution has added much complexity to the simple chordate body plan, but all of this rests ultimately on the ability of cells in the early stages of body building to receive and interpret signals coming from the notochord. We are, therefore, stuck with it. It may now be largely replaced, mechanically, by our complicated bony vertebral column, but we cannot do without its signalling early on. It remains an important part of our development, a 'living fossil' shorn of its old mechanical purpose but kept for just long enough to fulfil its old organizational responsibilities. Later, we break it up and use its remains to make intervertebral discs to cushion loads between our vertebrae[26] (these are the things that, when damaged, give rise to the painful condition of a 'slipped disc').

As if giving the body its main axis and creating the first tissue layers were not enough, the node/primitive streak system manages to perform yet another important function: it breaks the mirror-like symmetry of left and right.[27] It does this by moving fluids in a usefully inefficient manner. Many animal cells possess tiny, flexible, bristle-like protrusions called cilia. These are equipped with tiny motors, proteins that take energy from chemical reactions and use it to exert mechanical force on other proteins, and these motors cause the cilia to 'beat'. In primitive, single-celled animals, the beating of cilia acts as a propulsion mechanism to move them through a fluid. In humans, cilia on fixed cells are used in a similar way to move fluid: for example, cilia carried by cells lining the airways clear the lungs of mucus, while cilia carried by cells lining the oviduct waft eggs and early embryos down towards the uterus. Cells in the node make cilia too, and they project from the underside of the node into the fluid below.

The cilia made by the node have two unusual features. The first is that they stick out from the cell at an angle of about 45 degrees, pointing downwards and rearwards—the rearwards slant is set by the cells involved being sensitive to the head–tail polarity of the whole embryo.[28] The second is that they do not beat like a whip but instead show a circular motion, reminiscent of a cowboy twirling a lasso just before throwing it. This circling is rapid, running at about six hundred revolutions per minute (the idling speed of a typical car engine). The rotation is always clockwise, when viewed looking up the cilium towards the cell, because the motor protein complexes are 'handed' and can only attach to and turn the cilium one way.[29] The position of each cilium means that when it is

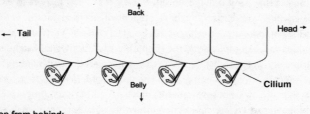

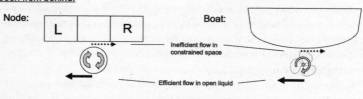

So net fluid flow is to left

FIGURE 14 Cilia rotating near a surface create a left-biased flow. The top panel shows a side view of the node, with cilia projecting down at about 45 degrees towards the future belly side of the embryo and pointing tailwards. The cilia beat clockwise, their path being shown by the cones. The bottom left panel shows a view from the tail end of the embryo, and illustrates how the position of the rotating cilia creates an imbalance in fluid flow, pumping more efficiently to the left. The bottom right panel shows a view from astern of a boat, and depicts an analogous effect familiar to those who steer boats in tight spaces.

at the bottom of its circle it is moving to the embryo's left, and when it is at the top it is moving to the embryo's right (Figure 14). So far, so symmetrical, but the closeness of the cilium to the cell at the top of its circle, when it is moving to the right, means that the fluid it moves is seriously slowed down by viscous drag against the cell's surface. When the cilium is at the bottom of its circle, it is far from the cell surface, effectively free of surface drag effects and able to push fluid efficiently. There will therefore be an imbalance between the amount that the fluid is pushed left and the amount that it is pushed right. This is analogous to the 'prop-walk' familiar to skippers of single-engine boats, where the limited space between propeller and hull creates an asymmetry of thrust that can make life so interesting when manoeuvring in a crowded harbour (Figure 14).

The biased flow created by the cilia means that the left side of the node is constantly bathed with fresh fluid, drawn up from the bulk liquid below the cilia.[29] Cells there can take up lots of small substances like calcium from it, while the right side of the node is left with exhausted fluid from which substances have already been taken. Any proteins released by the cells of the node will also be swept leftwards.

Cells of the node release various proteins into the fluid below them. These include Nodal, a powerful signalling molecule named after its site of origin. Both sides of the node make Nodal to some extent but, once it is released, it gets swept leftwards by fluid flow. Production of Nodal is increased by calcium, so the fresh, calcium-rich fluid coming into the left side of the node means that cells there make more Nodal than cells on the other side. Nodal affects the production of other proteins, including those that affect gene expression, so its building up on the left side of the embryo results in that side activating a slightly different set of genes to those on the right. The mirror image left–right symmetry of the embryo is therefore broken.

The loss of perfect mirror symmetry is useful for the way that we build our bodies. Most of our externally observable anatomy is mirror-symmetric across the left–right axis, but many of our internal structures are located asymmetrically. The heart and its circulation are asymmetrical, we have a spleen and pancreas mainly on the left and liver and appendix mainly on the right, and we have a host of subtle left–right differences in the brain. Some asymmetries can be seen even from the outside. In men, for example, one testis (the left in about two-thirds of men) hangs lower than the other.** The ability to develop asymmetrically may not have been absolutely essential to the evolution of advanced vertebrates, but it almost certainly made it much easier. The alternative would probably require us to have two of all organs except for gut, central nervous system, phallus, vagina, and bladder (all of which form along the midline anyway); this would create problems of packing and, while it may suit a long, slender body like that of a fish, in which the pairs of organs could be arranged one after the other, it would probably not be as useful to a large land animal that has to carry its own weight or fly.

What is remarkable about this mechanism for breaking left–right mirror symmetry is that the asymmetry on the scale of the whole body derives ultimately from asymmetry on the molecular scale, in the protein complexes that drive motion in the cilia. It is one of the few cases in which a property of a molecule is translated fairly directly into a related property of a whole body. The

---

** For his very detailed study on the representation of scrotal asymmetry in Greek sculpture, published in 1979, I. C. McManus was awarded the 2002 Ig-Nobel Prize for Biology by the journal *Annals of Improbable Research*, a journal devoted to research that makes people laugh and then think.

mechanism is unusual, but its existence is now supported by very strong lines of evidence. First, there is direct observation of the spinning of the cilia. Then there is modelling of the flow they would be expected to create, first mathematically and then by making a tiny array of working artificial cilia.[30] There has been direct visualization of the fluid flow using small particles dropped into the area, in both embryos and models, and there has been a great deal of measurement of Nodal production and accumulation. The story is also supported by a range of mutants that cannot make cilia, or whose cilia cannot move: these build bodies with random left–right orientation. There is one mouse mutant, called *inv*, that reverses the way that cilia point, so that they are at 45 degrees towards the head, rather than towards the tail. The net fluid flow is therefore to the right, and theory predicts that animals carrying this mutation will show complete, predictable left–right reversal of their body plan. This is exactly what is seen. Some humans are born with a similar reversal, presumably for the same reason.

The events in this chapter take only two or three days (days 15–17 from conception), yet they change the nature of the embryo utterly. Before, there was just a simple, featureless disc that showed no hint, in its shape, that it was anything to do with a complex animal. By the end, there is an elongated body, with clear head–tail, back–belly and left–right directions, with three distinct tissue types arranged in a definite order, and with a central notochord running along the body. The basic form of an animal is there, and elaboration of internal structure can begin.

5

BEGINNING A BRAIN

The brain: an apparatus with which we think we think.
Ambrose Bierce

The nervous system of an adult human is an immensely complicated structure that connects to almost every other part of the body. It is centred on the spinal cord, a long tube about an inch in diameter that runs down the middle of the back (Figure 15). Within the head, the basic structure of the spinal cord is modified to produce a set of swellings that are known, collectively, as the brain. From the spinal cord, nerves emerge to send signals to the muscles of the body and to receive sensory information such as touch, stretch, and pain. In addition, there are also quasi-independent nervous systems that regulate the behaviour of the internal organs such as the gut and heart, most actions of which are not under conscious control. Even these, though, do receive some control inputs from the spinal cord.

The nervous system is very important in controlling the biology of the adult body—so important that most countries regard death of the nervous system, 'brain death', as an indication of the legal end of life, even when the rest of the body is being maintained by machines. It does not, however, play any significant part in controlling the life of the embryo until it has grown into quite a mature foetus. Nevertheless, the nervous system begins to develop long before most other internal organs. There are several possible reasons for this. It may be because the basic architecture of the nervous system, which spans a great deal of the body, has to develop before other organs get in its way. On the other hand, early development of the nervous system may occur because it is one of the

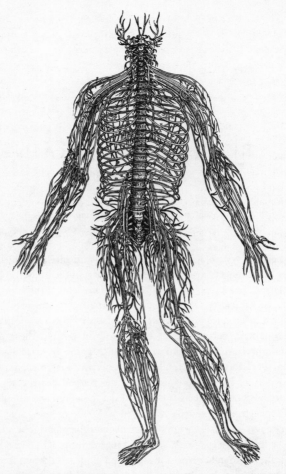

FIGURE 15 A drawing, by the Renaissance anatomist Andreas Vesalius, of the routes taken by the nerves that emanate from the spinal cord and connect it to the rest of the adult body. This drawing does not show the brain itself, which has few direct connections of its own, relying instead on the spinal cord to pass its messages to and from the body (the brain does connect directly to the retina of the eye and to the nose, both of which are, developmentally speaking, part of the brain anyway).

The Granger Collection/Topfoto

oldest parts of our body in evolutionary terms, being present in ancestors even more primitive than fish. As a sweeping generalization, to which there are a number of exceptions, the order in which new structures appear in the embryo follows the order in which they appeared over evolutionary time. Nobody knows for sure why this is so, but a popular theory is that tinkering with the basic developmental mechanisms that set up the body plan is more likely to make a horrible mess of the embryo than tinkering with mechanisms that add minor details to it. Random mutations that affect basic body plan will therefore probably lead to a failure to make a body at all, whereas mutations that affect only the later mechanisms are a little more likely to yield viable organisms, a few of which may be good at exploiting a new ecological niche and so ultimately develop as a new species. Over evolutionary time, new species seem to have arisen mainly by changing the later parts of their development, not the earlier, and embryos look much more alike than the adults that develop from them. It is interesting that the *very* early events of development, the ones that occupied Chapters 2 and 3, have been much more susceptible to evolutionary change, perhaps because they precede setting up the body plan so there is less potential for disaster when they are changed. Further reading on this topic can be found at the end of this book.

Whatever the reason, the nervous system does begin to develop very soon after gastrulation—indeed, it begins to form towards the head end of the embryo while gastrulation is still going on nearer the posterior. The nervous system—all of it—derives ultimately from ectoderm (the outer 'skin' of the embryo), specifically from a strip along the centre of the back from head to tail. The cells of this strip have to stop being ectoderm in contact with the outside of the embryo, and must instead form an internal tube which will go on to form the spinal cord and brain. What is more, they have to make this transition with out making a tear in the integrity of the embryo's surface. This is an example of the problem mentioned in Chapter 1: the embryo has to make large modifications to its structure without falling apart in the process. The solution involves a combination of cells changing their neighbours, stretching, and undergoing local changes of shape that cause the entire sheet to undergo a process of origami-like folding.

The first sign of nervous system formation is marked by an obvious change in the shape of the embryo. Gastrulation had already distorted the simple disc into a more oval shape, elongated along the future axis of the body; after gastrulation, this effect becomes even more pronounced, and the short, stubby

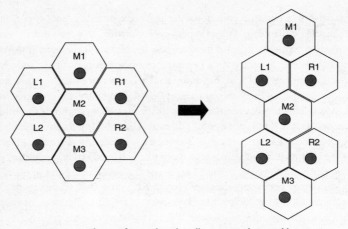

FIGURE 16 Change of tissue shape by cells swapping their neighbours.

embryo rapidly becomes long and thin, as cells continue to stream in from the sides, swapping neighbours so that they stack up more and more in the head-to-tail direction. Figure 16, which is drawn from a movie of an analogous process in fruitflies, shows one way in which cells swapping neighbours can alter the shape of a whole tissue.[1, 2] Cells that are initially separated, such as the cell labelled L1 and the cell labelled R1, move together to become new neighbours. At the same time, cells that were initially neighbours, such as the middle row labelled M1, M2, and M3, let go of one another and separate. In this way, a short, stubby tissue becomes long and thin.

As well as swapping of neighbours, there is also probably direct migration of cells towards the centre line. Both processes are guided by cells' abilities to read signals in the embryo (most of which remain to be identified, although we have a few clues[3]) that tell them which is the head–tail direction, and which the left–right direction. This internal 'compass' gives the cells a sense of direction within the plane of the cell sheet. The overall effect of the coupled left–right convergence and head–tail extension, guided by its compass,* is that the embryo begins to take on the elongated shape characteristic of the mature body. This means, in particular, that the previously broad and stubby region of cells along the mid-line of the back becomes long and thin, although it remains somewhat

---

* The 'compass' is planar cell polarity (PCP).

broader where the head will eventually form, so that the body looks a little like a keyhole.

With the basic shape of the body set up, neural development can begin in earnest. The notochord, which already runs along the centre-line of the body (Chapter 4), secretes signalling proteins. These cause any ectoderm within range, which in practice means the strip of ectoderm immediately overlying the notochord, to stop being run-of-the-mill ectoderm, and instead to prepare to make neural tissue. The cells here switch on genes that were previously silent; they thicken a little in preparation for shape changes that are to follow. Collectively, they are now known as the neural plate. Within the neural plate, cells become subtly different from each other according to their position: those immediately above the notochord becoming a central stripe and those near the borders with still-normal ectoderm making two edge stripes, in response to signals from that normal ectoderm. The stripes are not visibly different at this stage, but their positions will come to define the centre and the edges of a deep valley that is about to form (Figure 17).

The cells of the neural plate are all connected to one another by the same type of cell junctions that we met in the context of cells of the early embryo sticking

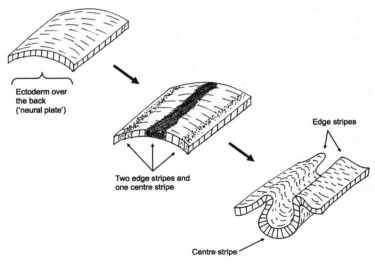

Ectoderm over
the back
('neural plate')

Edge stripes

Two edge stripes and
one centre stripe

Centre stripe

FIGURE 17 Formation of three 'stripes' along the ectoderm of the back (dorsal surface) of the embryo, which result in the formation of a valley.

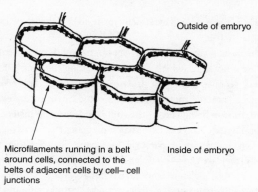

Outside of embryo

Microfilaments running in a belt
around cells, connected to the
belts of adjacent cells by cell–cell
junctions

Inside of embryo

FIGURE 18 The cell junction–microfilament system provides an integrated mechanical network that runs across the ectoderm and the neural plate. Note that the network is located closest
to the apical (outside) face of the cells.

to one another to form a compact mass (Chapter 3). Inside each cell, microfilaments of protein run from junction to junction, so that the complete junction–
microfilament system provides a continuous mechanical network (Figure 18).
Importantly, the junctions are located not around the middle of the cells, but
mainly near the surface that faces free space outside the body.

The existence of this mechanical network means that no cell can change its
shape without affecting the shapes of all cells around it. This means that the action
of individual cells can force the sheet to buckle. Cells in the centre stripe make
high levels of a protein (Shroom) that interacts with the microfilament system
and re-arranges it so that the junctions are drawn closer together.[4] Seen from the
side, the profile of each of these centre-stripe cells therefore changes from a rect-
angle to a wedge-shape (Figure 19a–b). The cell junctions remain tight, though, so
that, when the individual cells become wedge-shaped, it is impossible for spaces
to open up between them. Inevitably, therefore, the whole sheet of cells is forced
to curve[5] (Figure 19b). This curvature causes the centre line of the ectoderm along
the back of the embryo to fold inwards to make a deep valley (Figure 19c).
Conversely, the edge stripe cells seem to expand at their opposite ends by a mechanism that is not yet well understood; this causes the sheet to buckle the other
way, making a peak rather than a valley. Both actions have the same general
effect—they cause the region of ectoderm that will form neural tissue to dip down
deep into the embryo and the flanking ectoderm to move closer together.

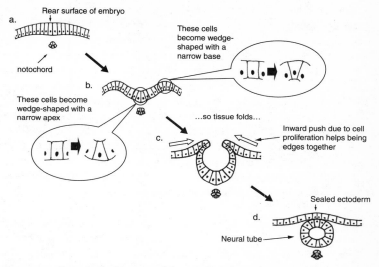

FIGURE 19 Stages in the formation of the neural tube. Each figure shows a cross-section through the embryo's back. Initially straight cells, in (a), become wedge-shaped, driving the formation of a valley (b–c), the sides of which eventually fuse to turn the valley into a tube (d).

The formation of the central valley illustrates how large-scale tissue organization can take place by strictly local forces. If the piece of ectoderm that will undergo this process is cut out of an animal embryo and kept alive in a culture dish, it still generates the characteristic valley on time, showing that the rest of the embryo is really not needed to control the process once it has started and it really is local to the cells involved. There the development stops, though, because the next stage depends on activity in the flanking ectoderm that will not itself form nervous tissue. The cells of this ectoderm do three important things: they flatten out a little, so that they become less tall but wider; they multiply; and they continue to converge on the centre line of the embryo as part of the thinning and lengthening process already described.[6] These three activities all create a sideways pressure that pushes the edges of the valley towards one another until they meet (Figure 19c–d).

When they meet, the tops of the two sides of the valley adhere to one another and therefore turn the valley into a tube that is still connected to the ectoderm, at least initially. As already-meeting regions of the neural tube fuse, they help to draw neighbouring regions together and the fusion therefore spreads along the top of the valley in a way reminiscent of a zip fastener. Once the two sides of the neural tube have met, the cells rearrange themselves so that the neural tube separates completely from the ectoderm that flanks it, without producing a hole in the embryo. The details of how it does this have not yet been investigated thoroughly, but a plausible mechanism, based on the cell adhesion molecules known to be present, might be as follows. The adhesion between cells of the neural tube, which operates using proteins such as N-cadherin, may be much stronger than the adhesion between these cells and the ectoderm that flanks them. The neural tube cells at the boundary therefore try to maximize their contact with other neural tube cells, which can be done only at the expense of contact with the flanking ectoderm. Once the ectoderm cells from opposite sides of the neural tube touch, the adhesion between them, mediated mainly by a molecule called E-cadherin, is stronger than the adhesion between them and the neural tube cells that are still attached to them. The ectodermal cells therefore increase contact with each other at the expense of their contact with the neural tube cells. This 'birds of a feather flocking together' requires no special mechanisms: it emerges from simple biophysics of adhesion. Eventually, the two tissues will simply lose contact with one another and the ectoderm will exist as a continuous 'skin' across the back. It will remain in place and give rise to the outer layer of the skin of the child. Note, though, that this mechanism is conjecture, and it is not even certain that like cells do adhere much better than unlike cells, in this system.

Closure of the neural tube is an error-prone process in humans and it fails in some individuals, at least along part of the spinal cord or brain. When closure fails along part of the spinal cord, a child is born with the condition spina bifida (split spine). In Scotland, where this book was written, the condition was once very common and affected nearly one in a hundred children conceived in some areas. In at least a quarter of cases, the underlying problem seems to be that the two sides of the neural tube never even meet because the ectoderm that flanks them fails to push them inwards properly. It turns out that closure is very sensitive to the amount of folic acid (vitamin B9) in the tissues, too little folic acid meaning too little proliferation and a high risk of spinal bifida. The word 'folic' means 'from leaves', and green vegetables, beans, and some fruits and seeds form the main natural dietary source of folic acid. Unfortunately, these essential foods are virtually missing from the diets of many industrial populations, the poorer areas of urban Scotland being particularly infamous in this regard.

For this reason, several groups of researchers have studied the effects of giving would-be mothers extra folic acid, in the form of vitamin supplements, from before conception to well after neural tube formation has been completed. In almost all published studies, the effect of the treatment was a dramatic reduction, by roughly one half to two-thirds, in the incidence of spina bifida.[7, 8]

The implications of these studies, that folic acid supplements could significantly decrease the incidence of a disabling developmental defect, were quickly translated into a recommendation that women contemplating pregnancy should take folic acid supplements. This simple advice does not help with unplanned pregnancies, though, because neural tube closure happens only three to four weeks after conception, at which time many mothers will not be aware that they are pregnant. Unfortunately, the incidence of unplanned pregnancies, and the probability that pregnancy would not be recognized early enough, are highest in precisely the economically and educationally poorest section of the population that is also least likely to eat fresh vegetables. For this reason, several countries, including the US, now insist that basic foodstuffs such as bread and breakfast cereals are enriched with the addition of folic acid so that the whole population is treated. Having a state insisting on the adulteration of everyone's food, even with a naturally occurring molecule that is believed to be safe, is ethically contentious. At the time of writing, no European Union country requires the addition of folic acid to foods, but foods such as breakfast cereals that are enriched in the molecule are readily available.

Spina bifida is generally compatible with life although, depending on the location and severity of the defect, a child may suffer from mild or serious disability, paralysis of the lower limbs and lack of bladder control being common. Where neural tube closure fails in the area of the brain, a much more serious condition called anencephaly results. The back of the head and much of the brain is missing and this is lethal, death often occurring before the child is born or, if not, then very soon afterwards. It should be noted that a lack of folic acid described above is not the only thing that can cause spina bifida and anencephaly: genetic and other environmental influences are also important, and even mothers who are very careful about their diet and lifestyle can give birth to children with this condition through no fault of their own.[9]

There is one other, very rare but remarkable, abnormality that can occur during the process of neural tube closure. In some types of twinning, two embryos can develop very close to one another but may be different sizes. Very occasionally, one small twin will be caught up in the valley of a larger twin's neural tube and will be taken up and sealed, forever, inside.[10] There it can continue to develop either as a disorganized tumour or as a tiny, basically normal foetus. One of the clinical reports listed in the technical references, for example, describes a six-week-old baby whose head was expanding unusually; inside one of the normal cavities of the baby's brain was the tiny body of his twin, complete with recognizable limbs, trunk and head.[11] This phenomenon is one of a class called *fetus in fetu* (literally 'a foetus inside [another] foetus'). It is found more often, though still with an incidence of only about two per million births, following another normal sealing-up event that takes place in the belly, rather than in the head and later in development. The presence of this abdominal *fetus in fetu* is usually detected within weeks or months of a baby being born, but sometimes it can escape detection for a very long time. Another report[12] describes the case of a man who was discovered to have been carrying his unknown twin inside his own abdomen for thirty-nine years.

These two abnormalities of neural tube formation, the relatively frequent spina bifida and the very uncommon neural *fetus in fetu*, have in common the fact that they are caused, at least in part, by factors outside the embryo itself. Spina bifida is caused, or at least made more likely, by poor maternal nutrition, and *fetus in fetu* by one embryo happening to be just where another embryo is about to close over. They underline the important point that embryonic development is not a deterministic outcome of genetics alone, but is the result of an interaction between genes and environment, even deep inside the womb.

The neural tube, formation of which has been the focus of this whole chapter, does not yet contain any actual nerve cells or neural connections. It harbours no sensations, no reflexes, no mind, no will, no thought—they all come much later. For this reason, the young neural tube is not capable of performing any of the command-and-control functions that the brain and spinal cord will one day assume. It has, though, produced their basic plan and it is poised to send out cells that will build the other, outlying parts of the nervous system very soon; how it does that will be the topic of Chapter 7.

The general elongation of the body that has been a background to all of the events in this chapter plays an important part in forming another tube, that of the gut. Gastrulation (Chapter 4) creates the endoderm layer that will one day make the gut, but at this stage the endoderm is just a flat plate at the front (ventral) surface of the embryo, facing the open space of the yolk sac (Figure 20a).

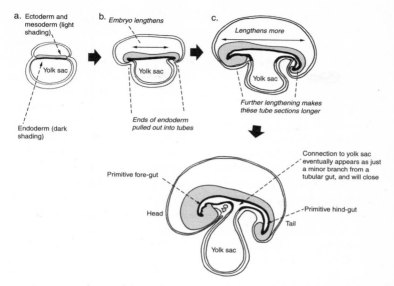

FIGURE 20 Lengthening of the body, which accompanies the formation of the neural tube, is also important for formation of a gut tube from an initially flat plate of endoderm. As the embryo elongates, it pulls the endoderm into a tube at each end of the body. As the embryo grows still more, the part of the endoderm that still faces the yolk sac becomes a proportionately smaller and smaller part of its total area, and the gut is essentially almost all a tube now. By the stage of the last diagram, the gut tube will have acquired various new branches, but these have been omitted for clarity.

Elongation of the body pulls on the ends of the endoderm, causing them to 'overshoot' the entrance to the yolk sac and to be drawn out to make short tube-like extensions into the growing head and tail ends of the embryo. As the body grows ever longer, these tubes become longer with it, becoming recognizable as the primitive fore-gut, from which the oesophagus, stomach, and upper intestines will form, and the primitive hind-gut, from which the lower intestines will form (Figure 20b). The connection to the yolk sac, which seemed huge at the scale of an embryo just-gastrulated, fails to keep up with general embryonic growth so that it soon has the appearance of only a minor branch from the ever-enlarging gut of an ever-enlarging foetus (Figure 20c). This branch is eventually closed off, leaving the gut as a tube fully enclosed within the body (except where its ends make new connections with the outside, at the mouth and anus).

The stage of development reached at the end of this chapter, an elongated embryo with three main tissue layers, and with a notochord, a dorsal neural tube, and a ventral gut tube (albeit still with a yolk sac connection), already has most of the basic structure of a chordate animal. It is certainly a very long way from the end of development, but it has arguably reached the end of the beginning. In less than four weeks, a single-celled embryo has organized itself into a collection of thousands of cells arranged as a simple body. It has pulled itself up by its own bootstraps, first using simple geometrical tricks to create differences where there were none, and then using these new differences between cell types as a source of information to create yet more differences, more information, and more organization. All of this was accomplished by relatively simple mechanisms that obey relatively simple, local rules. Having reached a stage of modest complexity, the embryo has enough internal information to be able to build far more complexity fairly quickly. Its anatomical form will change a great deal but, as we will see in the next section, the underlying essence of what it does, in the sense of using simple rules and local interactions, will not change at all.

6

LONG DIVISION

Nothing is particularly hard if you divide it into small jobs.
Henry Ford

An important feature of bodies that has hardly been mentioned so far is that they are not the same all the way along. Some features, such as the presence of the neural tube, extend all the way from the head to the posterior; many features, however, appear only at particular levels along that axis. Arms, for example, protrude from the body at a level about three-eighths of the way along the head–tail axis, in an adult, and legs protrude from almost the end of that axis. The trunk between the arms and legs is devoid of limbs. Internally, there are similar unique locations of organs: eyes are to be found an eighth of the way from head to tail, while the liver is to be found at a site about six-eighths of the way.* Each of these special structures develops from the cells of the embryo, and the appearance of different structures at different head–tail positions implies that cells of the embryo must be different at different points along the head–tail axis. This chapter is about how these differences come to be. To understand the mechanisms involved, it is useful to understand an important but almost hidden feature of human anatomy.

When one looks at the structure of simple garden invertebrates such as earthworms, centipedes, or wasps, it is obvious that their body plan consists of

---

* Any artists surprised by the proportions quoted here should note that they follow a zoologist's convention of using the head–tail length as the length of the body, and not adding in the length of the legs.

a series of repeated segments. An earthworm, for example, consists of a long cylinder with many circumferential lines around it, each of which marks the boundary between two adjacent segments (Figure 21a). Most of these segments are identical to the naked eye, although the segments at the head, tail, and reproductive areas do look a little different. A centipede is similar in form although its body segments each carry a pair of legs. Again, the head and tail segments are rather different. The abdomen of a wasp is also obviously segmented, the coloured stripes emphasizing this. With a hand-lens, it is clear that the small thorax ahead of that abdomen also consists of three segments, although these look rather different from each other; each carries a pair of legs but only two segments also carry a pair of wings (you may need to look carefully to see that wasps have four wings; the rear pair are small, and are held very close to the trailing edges of the front pair).

The trick of dividing a body into many segments, which are basically similar but carry certain specializations according to how far along the body they are, is a feature of many types of animal. In primitive species such as earthworms, each segment is remarkably self-sufficient in physiological terms, having its own primitive lungs, kidneys, nervous system, etc. Some systems, such as the gut, run through all or most of the segments, and the nervous systems of each segment are connected so that the animal behaves as a coherent whole. Nevertheless, building much of the body by repeating a basic module probably confers a great advantage in being able to use the same genetic and cellular systems again and again. This economy will have made the evolution of animals of this scale much less unlikely than it would have been had every part of the body been radically different. It is in the most primitive animals, primitive both in the sense of simplicity and in the sense of early appearance in the fossil record, that segments are most similar to one another. During subsequent evolutionary history, there has been a gradual pattern of making particular segments more and more specialized for particular tasks, such as powering insect wings or walking. As segments have become more specialized, they have become more dependent on each other physiologically so that, instead of each segment having its own kidney, breathing apparatus, etc., they share centralized services.

Vertebrates are segmented too, although this is seldom obvious at first glance. Fish, for example, look smooth on the outside and give few clues to having a segmented anatomy, but anyone who has eaten a fish will have noticed how most of its body consists of a repeated series of muscles (the 'meat' of the fish) arranged along a backbone made of a repeated series of vertebrae and fine ribs. Humans, whose evolutionary line includes fish-like ancestors, have a similarly segmented anatomy, although it is even less easy to discern from the outside. The easiest place to see the segmented nature of humans is in the skeleton, specifically the backbone (Figure 21b). This consists of thirty-three vertebrae, each of which is a variation on a common theme. The top two, in the neck, are specialized to allow their possessor to nod and shake his head. There are five more in the neck, then twelve in the chest region that connect to ribs. Below the chest are five vertebrae of the lumbar region, which are built heavily for weight bearing. Below them are nine more small ones that fuse together to make the sacrum and coccyx ('tail bone'). The soft tissues that attach to these bones also look segmented: for example, the muscles that connect adjacent ribs and the nerves that make connections between the body and the spinal cord. Other structures of the body show segmentation as well, but mostly in a more subtle way.

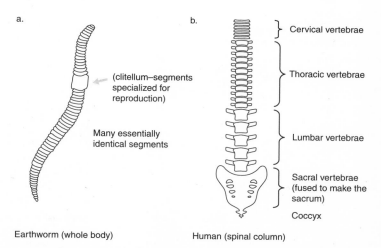

a.

(clitellum–segments specialized for reproduction)

Many essentially identical segments

b.

Cervical vertebrae

Thoracic vertebrae

Lumbar vertebrae

Sacral vertebrae (fused to make the sacrum)

Coccyx

Earthworm (whole body)   Human (spinal column)

FIGURE 21 Segmentation in (a) the earthworm, an animal that is obviously made of a series of repeating segments, and (b) in the human, which is less obviously so. The illustration of the human spinal column omits appendages such as limbs and ribs, and also omits the head and its articulation (including the highest vertebra), in order to avoid distractions from the segmented nature of the spinal column, which is the ancient core of the skeleton. Also, the vertebral column drawn has been slightly stylized: vertebrae of each type show minor detailed differences from each other that have not been depicted here.

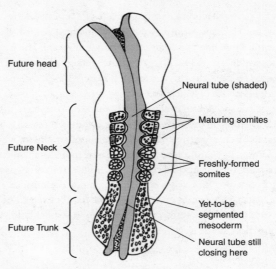

Future head

Neural tube (shaded)

Maturing somites

Future Neck

Freshly-formed somites

Yet-to-be segmented mesoderm

Future Trunk

Neural tube still closing here

FIGURE 22 Somite formation in a human embryo, at about five weeks; six somites have formed so far and these will give rise to the base of the skull and to the first vertebrae of the neck. The embryo will continue to extend towards the bottom of the page as it develops.

During development, the visible segmentation of the human body begins in the mesoderm that lies to the left and right of the neural tube. The cells in this tissue, produced in gastrulation (Chapter 4) are initially associated with one another only loosely and are not organized in any special way. During the course of segmentation, however, this tissue becomes divided into a series of blocks called somites; these are the precursors to vertebrae, to muscles of the trunk, and to a few other body parts. The formation of these somites does not happen all at once but rather in a sequence from the neck of the embryo towards the tail, so that pairs of somites appear one-by-one (pairs, because a somite is added on the left and right sides of the neural tube at the same time). In humans, a new somite forms every six hours or so (Figure 22).

The task of turning a section of unsegmented tissue into a new somite is achieved by cells altering their adhesive properties. These cells, which were highly adhesive cells in the ectoderm before gastrulation and became loose mesoderm during gastrulation, now re-activate their adhesion systems.[1]

Once they have done this, they separate from their old mesodermal neighbours and organize themselves into a continuous, closed, cyst-like sphere that is the new somite.

It is clear that the whole of the unsegmented mesoderm cannot make this switch to adhesion at the same time—if it did all that would result would be one large 'mega-somite' instead of the series of small individual somites that is actually required. Instead, the embryo has to use some mechanism for telling just short blocks of cells to change their adhesion at any one time, and to let one somite finish forming before the next one starts. This requires cells to detect, very accurately, whether they are in the place that a somite is going to be.

In earlier stages of development, the embryo was still very small when cells had to respond differently according to where they were, for example to make a primitive streak. At gastrulation, for example, the embryo was small enough for the concentration gradient of a signal from the AVE to impose a crude head–tail axis. As the embryo becomes larger, however, simple concentration gradients like this become a less and less suitable mechanism for imposing body-scale patterns. The longer a tissue that is to be patterned by a single concentration gradient is, the less steep the slope of that gradient will be compared to the size of any cell (Figure 23). The shallower the gradient, the less the difference in concentration between adjacent cells, and the harder it would be for cells to make

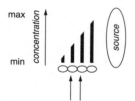
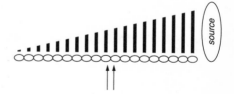

Gradient along a short tissue: adjacent cells have a clear difference of level so their behaviour can be reliably different at different places along the gradient.

Gradient along a long tissue: adjacent cells have only a small difference of level, risking serious errors in the relationship between position and behaviour.

FIGURE 23 Concentration gradients spreading from a source work well to specify positions over distances that are fairly short, but as the distances get longer, adjacent cells experience smaller differences in concentration and, given the inherent noise in biological systems, it becomes harder for cells to determine whether they are just inside or just outside the space in which their behaviour should alter in a particular way. Fixed gradients are therefore not suitable for imposing rich patterns, with many possible behaviours, on large tissues, and cannot be used for patterning all of the somite boundaries along an embryo.

accurate decisions that are meant to reflect their exact positions in an embryo. The gradient system that was adequate for imposing crude differences in cell behaviour in the short early embryo is simply not up to the task of imposing a succession of fine differences along the much larger body that now exists.

The embryo solves this problem, not by discarding the idea of gradients entirely, but rather by using short-range gradients to specify a somite, and then moving the gradients one somite-length along the embryo and using them again. The mechanism that achieves this seems almost fanciful in its complexity, but it has been verified by a large series of careful experiments in a range of species and, while we certainly have more to learn about it, we know enough to know that the mechanism is quite real.

All the way through the process, the cells of the mesoderm that has not yet been segmented into somites, and therefore lies tailward from any already-made somites, make a signalling protein called FGF.[2] This is their default activity, and each of the cells will continue to make FGF until something stops them. The cells can detect the FGF they themselves make, and it prevents them from making the change to their adhesive systems that would be needed to become a somite. In other words, the not-yet-segmented mesoderm keeps telling itself 'not yet', in the language of FGF.[3]

Cells that are already in the process of making a somite do not make FGF. Instead, they make another signalling molecule, retinoic acid, which is also made by cells in the head of the embryo. Retinoic acid spreads from the last somite into the nearby, not-yet-segmented mesoderm. It cannot spread very far, because this mesoderm makes a molecule that destroys it, but it can penetrate at significant concentrations for around a tenth of a millimetre.[4] Retinoic acid switches off production of FGF in this mesoderm. In this way, it creates a narrow 'window', or 'permissive zone', of not-yet-segmented mesoderm in which there is not enough FGF to prevent cells making a somite (Figure 24).

This definition of a window in space is one important part of the mechanism that produces new somites one by one, but it is not enough to solve the problem. After all, if cells could commit to being somites as soon as the permissive zone has moved far enough tailward to include them, then cells would make this commitment one after the other and there would never be a clear posterior end to each somite, because there would always be extra cells just released from inhibition, trying to say *me too!* Such a system would fail to divide the tissue into blocks, and all that would result would be the building of one very long, fuzzy-ended 'mega-somite'. The problem could, however, be solved if cells were to

(Head end)

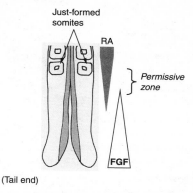

FIGURE 24 The opposing gradients of retinoic acid (RA) from the head end of the body, including maturing somites, and of FGF from the tail end, define a narrow 'permissive zone' within which cells are capable of committing themselves to make a new somite.

desist from committing to make a somite until the permissive zone had moved about a somite's length more tailwards.[5] All of the cells in the permissive zone at this point would then be allowed to commit but no more cell commitments would be allowed until the permissive zone had moved on another somite's length, and so on. This seems to be exactly how the system is arranged in real life: the switch between cells in the permissive zone being able to commit or not is being regulated by the 'ticking' of a molecular clock.

The 'clock' consists of a network of proteins that control each other's synthesis; in some cases they also repress their own synthesis, either directly or indirectly. To understand how controlled protein synthesis can make a clock, it is worth considering a very simple 'toy' system, much simpler than the real somite clock: this 'toy' system consists of a single gene that, when active, specifies the production of a protein that inactivates that very gene. In other words, the protein suppresses its own synthesis. We will assume that the gene is in some kind of cell, so that the basic biochemistry of making proteins can be taken for granted. Recall that active genes are first transcribed to make an RNA 'copy' of the gene, and that this RNA is then translated to make the actual protein (Chapter 1).

If we first assume that the processes of transcription and translation take very little time compared to the lifetimes of the RNA and protein molecules

A protein repressing its own synthesis with negligible delays in the system

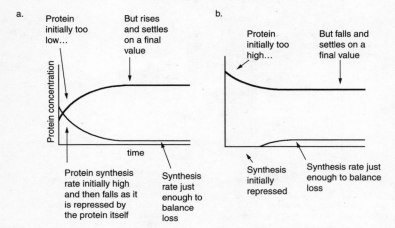

FIGURE 25 Behaviour of a system in which a protein (bold line) represses its synthesis of its own RNA (fine line), with a delay in response that is negligible compared to the life of the protein. Whether the protein concentration begins too low (a) or too high (b), it settles down on a constant value. Contrast this with the behaviour in the next figure, Figure 26.

once they have been made, then our 'toy' system would clamp the levels of its protein to a particular level. If the system began with less protein than this level, for example, there would be little repression of the gene's activity. RNA would be made and translated, and the level of the protein would therefore rise. There would be more protein molecules to inhibit the activity of the gene and new protein production would fall until the small amount of residual production exactly balanced loss (Figure 25a). If the protein level were initially too high, on the other hand, it would repress production strongly and no more of the protein would be made until the levels fell again as the proteins slowly aged and degraded; again, the system would settle to the point at which production just balanced loss (Figure 25b). The net effect of such a system would therefore be to clamp the levels of the protein within tight limits. For proteins that are very long-lived (hours or days) compared to the delays involved in transcription and translation (minutes), the small delays do not much matter and systems like this are actually used by cells to keep the amounts of particular molecules within tight limits.

A protein repressing its synthesis of its own RNA with large delays in the system

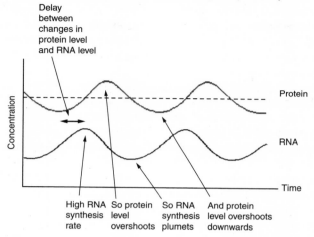

FIGURE 26 Behaviour of a system in which a protein represses the synthesis of its own RNA, with a delay in the system that is significant compared to the life of the protein. At the beginning of the time illustrated, the concentration of protein is falling from a high level but the concentration of RNA is low. The concentration of the short-lived protein continues to fall but, although there is now too little to inhibit synthesis of the RNA greatly, it takes a while for RNA to accumulate. The protein therefore falls away too much, and undershoots its average line. This undershoot allows the RNA to be synthesized very strongly, and soon a great deal accumulates. This means that protein levels rise but, with so much RNA around, the concentration of protein overshoots before its suppression of new RNA synthesis makes a significant difference to the concentration of RNA. This pattern of under- and over-shooting repeats, making a simple oscillator or 'clock'.

Now, if we alter our assumption and instead assume that the protein is so short lived that the time taken to make it is comparable to its lifetime, very different behaviour will be seen. Consider a situation in which our 'toy' begins with little of the protein. With little protein to repress synthesis, transcription of the gene can proceed quickly and, after a delay, levels of RNA rise and a little later levels of the protein itself rise as the RNA is translated. As the protein levels rise, transcription is reduced but there is still plenty of RNA so, until this RNA degrades, more protein is still made by translating this RNA. The level of protein therefore overshoots (Figure 26). The really high levels are enough to shut off transcription almost completely. As the RNA and protein degrade, the levels of protein fall and transcription starts again but, because of the delays involved in transcription and translation, it takes a while before the levels of protein can rise, and the continuing fast-degradation of protein means that the level now overshoots in the downwards direction. The system has too little protein again, and the cycle will begin again. The delays cause the system to overshoot repeatedly from too high to too low, and the result is not stability but an alternation of high and low protein levels that follow each other, like the tick-tock of a clock.

The network involved in the real somite clock is a lot more complicated than just one protein that controls its own synthesis,[6] but the core of the clock seems to be based on self-repressing loops of the kind described above that are reliant on proteins being short-lived compared to transcriptional and translational delays involved in their production. The large number of other proteins involved seem to be there to help the clock run very regularly and to keep the clocks of adjacent cells in step. In mice, the clock takes about two hours to complete a cycle. One phase of the cycle gives permission to any cells that find themselves in the permissive (low FGF) zone of the not-yet-segmented mesoderm to commit to being a somite. The alternation between the short 'tick' phase of the clock, when cells in the low FGF window are allowed to commit, and the long 'tock' phase when they are not allowed to commit even if they find themselves in the permissive window, means that there is time for the permissive zone to advance tailward by a somite length before the cells that find themselves inside it can commit. The mesenchyme is therefore turned into blocks defined by the length of the permissive zone and the speed of the clock (Figure 27).

The mechanism outlined above has an interesting implication: if the clock were to run faster, the permissive zone would advance less far tailward and a larger number of smaller somites would result. Some animals, such as snakes, have many more vertebrae than humans, because snake embryos make many more somites. Researchers have recently examined the somite clock of snake embryos and find that it runs very much faster than the somite clocks of mice and humans, just as the theory outlined above would demand.[7]

Four mechanisms that operate on the local, cell-scale level, then—production of retinoic acid, production of FGF, the somite clock, and tailward growth of the embryo—integrate space and time in a way that lets cells decide whether or not to participate in making a new somite. This happens without the cells having to have any internal map of the embryo or where they are within it. Again, the cells can organize themselves in a regular, large-scale pattern based on only local rules.

The system described above solves the problem of dividing mesenchyme into a series of segmental units. It does not, in itself, solve the extra problem

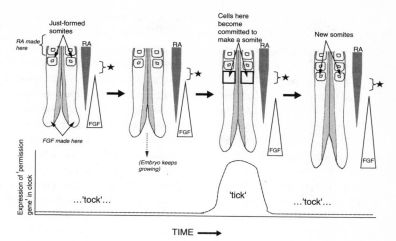

FIGURE 27 A current model of how both time and space are used to divide a continuous tissue into somites. The left-hand diagram, similar to Figure 24, shows the posterior part of the embryo just after a new pair of somites has formed. Retinoic acid ('RA'), made by the older somites, spreads tailwards and peters out gradually: this is depicted in the diagram by the shaded triangle. FGF, made in the tailbud, spreads headwards and also peters out gradually: this is depicted by the open triangle. The ratio of RA to FGF that cells experience is therefore different at different locations along the head–tail axis of the mesenchyme below the last somites. If cells find themselves within a particular range of RA:FGF ratios, the 'permissive zone' shown by the black star in the diagram, they will be able to commit themselves to making a somite. As time proceeds, the embryo continues to grow at the tail end and maturing somites start to make RA, so the critical window that defines cells capable of making a new somite moves gradually tailwards. The cells within this window will only actually commit to making a somite when the clock is in its 'tick' phase. By the time the next 'tick' is happening, the cells that were allowed to make a somite by this tick will have done so, so a clear boundary exists between the old somite and the new one. The process then repeats enough times to lay down all of the somites of the body.

that, in humans, the segmental units are not all the same but are specialized to form different types of vertebra etc. Somites in the thoracic region, for example, will produce medium-sized vertebrae that bear ribs, those in the lumbar region will produce large vertebrae that have no ribs, and those in the sacral region will produce small vertebrae that fuse together to make the back of the ring of bone that protects the pelvic region. The cells within each somite therefore have to 'know' which region of the body they are in. After several decades of intense genetic research, scientists are starting to understand how these cells know where along the body they lie. According to our current understanding (which, like our understanding of the somite clock, may well change as more is discovered), this system again operates by using time to transfer space information from the scale of molecules to the scale of the whole embryo.

The required spatial information is held at a molecular scale on DNA itself, as four clusters of genes. The genes, called HOX genes, come in thirteen basic types, numbered 1 to 13. The four clusters are called the HOXA, HOXB, HOXC, and HOXD gene clusters, and each contains a set of HOX genes arranged in numerical order (Figure 28). The genes are named so that the type 1 gene in the HOXA cluster is called HOXA1, the type 1 gene in the HOXB cluster HOXB1, etc. The HOXA cluster therefore contains genes in the order HOXA1, A2, A3, etc. No one cluster contains a version of every gene from 1 to 13 so that, if the positions of the genes are drawn out in a diagram that places genes of the same type in one column, each HOX cluster has gaps (Figure 28). There is very strong evidence, from other animals, that our remote ancestors had just one HOX cluster (as insects such as the flour beetle *Tribolium* still have) but that, during the evolution of vertebrates, the HOX cluster was copied twice to yield first two clusters, as found in jawless vertebrates, and then four clusters as found in jawed vertebrates.[8] That would explain why HOXA1 is so similar to HOXB1, HOXA2 is so similar to HOXB2, etc. Since that time, each cluster has lost some of its genes, presumably because, soon after the duplication took place, the genes were so similar that any of one type could deputize for the others. Loss of the hypothetical ancestral HOXD6, for example, would not have mattered if HOXA6, HOXB6, and HOXC6 could between them do everything that HOXD6 was doing. As the different genes of the same type started to mutate and diverge over long periods of time, they would have begun to acquire subtly different functions and further losses would have been impossible. Whether or not one

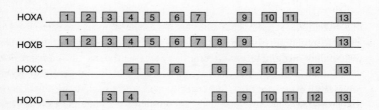

This ends comes on first   This end comes on later

FIGURE 28 The genetic structure of the four human HOX clusters. Each horizontal line represents a continuous region of a chromosome and each numbered box a HOX gene (box 1 on the HOXA cluster being HOXA1, etc.).

believes such evolutionary just-so stories, the fact is that the HOX clusters of humans do now have the form depicted in Figure 28.

It is striking that, during all of the 460 million years or so that jawed vertebrates have swum, walked, or flown this planet, the order of HOX genes along the clusters has never changed in any of these animals. In contrast, most other genes have seen many changes of relative position, which is reflected in their order being different in different types of animal today. The reason that the order of the HOX genes has been so conserved is that the order is linked fairly directly to the order in which different specializations of segments appear along the body.

The HOX gene clusters begin their body-patterning activity during gastrulation. At each position of the body, along the neck–posterior axis, the patterning activity takes place shortly before somites are formed. At the level of the embryo that will become the neck, as cells undergo gastrulation movements through the node, they activate genes from the left-hand end of the HOX clusters ('left' in terms of Figure 28). Later, as cells destined to make more posterior parts of the body become involved in gastrulation, genes from a little further right in the HOX clusters become activated. Later still, as gastrulation has reached well along the head–tail axis, genes well to the right of Figure 28 are active. In each case, cells remember the set of HOX genes that they switched on as they emerged from the node, and continue to have these genes active long term (that statement is an over-simplification, but it will do for the purposes of this chapter).

In summary, cells that gastrulate early switch on HOX genes to the left of Figure 28, while ones that gastrulate later switch on HOX genes further to the right. How might this be controlled? One possible method is to arrange for a wave of *potential* gene activation to pass slowly rightwards through the clusters of HOX genes while cells waited to gastrulate. According to this idea, when a cell dives through the node to undergo gastrulation, it *actually* activates the expression of the set of HOX genes that already have the potential to be activated at the time, and it remembers this set. Cells that gastrulate later and later, and therefore further and further from the future head of the embryo, will naturally therefore express a set of genes further and further to the right of Figure 28, because the wave of potential gene activation will have had more time to proceed rightwards through the cluster. By this mechanism, the order of genes in the HOX clusters, which is a structure at the molecular scale, is translated into the order of HOX gene expression along the embryo, which has already reached the millimetre scale. This is one of a very few examples in which genetic structure relates directly to embryonic structure.

Before exploring the consequences of different levels of the head–tail axis expressing different HOX genes, it is worth spending a little time exploring how it is that a wave of potential gene activation sweeps along the HOX clusters in the first place. The details are still the subject of intense research, but the skeleton of a mechanism does seem to be emerging, dimly, through the haze of thousands of detailed experimental results. Within chromosomes, DNA can be packed around proteins in a very compact form, useful for saving space but not useful for making the genes available to be transcribed. Alternatively, DNA can be in a loose, easily accessible form. Most of the time, most chromosomes have many stretches that are tightly packed, and also many looser loops of DNA coming out. Some stretches are always tightly packed. Some are almost never tightly packed, and some can be in either state, depending on what sequence-specific DNA binding proteins are present. Recent data obtained from experiments on mouse embryos suggest that at least one of the hox[†] clusters, hoxb, can be in either state.

Before the hoxb genes are activated, the whole cluster is in a compact state. Of the genes in the hoxb cluster, the right-most gene (hoxb13) is the one buried deepest in the compact coil and hoxb1 is the most accessible, being at the beginning of the compact zone. Expression of the hoxb genes depends on their being liberated from their prison of compact organization. This can be achieved by sequence-specific DNA binding proteins binding to the DNA at the most accessible end of this rather inaccessible unit and altering the packing proteins so that they allow DNA to adopt the more relaxed, looped-out state. The sequence-specific DNA binding proteins that are responsible for this are activated by retinoic acid. The effect of retinoic acid on the hoxb cluster is so powerful that it can be seen even in mouse embryonic stem cells grown in a simple culture dish.[9] If a tiny amount of retinoic acid is added to the dish then, bit by bit, the tightly-compacted hoxb cluster starts to loosen up from its end so that first hoxb1, then hoxb2, then hoxb3, and so on, become accessible to transcriptional machinery (Figure 29). All of this takes time, which is not surprising considering that the DNA of the hoxb cluster is about 150,000 bases long. The liberation of the hoxb genes is therefore a gradual process and one that has to follow the order in which they lie along the cluster. Presumably, the same is true[10] of the hoxa, c, and d clusters of the mouse and, by analogy, all of the HOX clusters of humans.

[†] By convention, the human genes are written all in capitals, HOXB, while the mouse genes are written as hoxb.

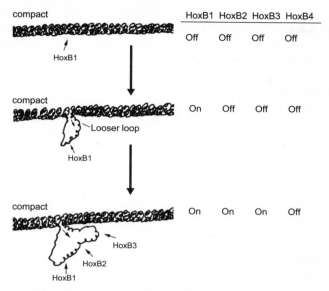

FIGURE 29 The gradual change of organization of the HOX cluster from its compact to looser form makes more and more of the genes available for transcription, in the order in which they lie in the cluster.

This opening out of the hoxb cluster into an expressible loop has also been observed in the mouse embryo itself, and the speed of the unravelling, and the wave of gene activation that it allows, fits well with the speed with which gastrulation reaches ever more tailward regions.[11] There is therefore a reasonable hope that the mechanism described above is correct, although we already know that it is surrounded by many extra complexities to do with making sure that cells remember exactly what they expressed, with making neighbouring cells agree on what they are expressing, and with shutting down the left-most genes when ones further to the right are expressed. It will no doubt take a large fraction of the careers of many researchers to sort all of this out properly.

The human HOX genes (and the mouse hox genes) are important because there is a link between the set of HOX genes expressed by a somite and the type of vertebra it makes.[12] The third, fourth, and fifth vertebrae of the neck, for example, are very similar and the cells that make them all express the same set of HOX genes. The next two vertebrae along the body are different and the cells that make them express subtly different sets of HOX genes. Different species of animal can have different numbers of vertebrae, chickens having fourteen neck vertebrae, and mice, like us, having seven. In both animals, however, the beginning of expression of HOX6-type HOX genes (Hoxa6, Hoxb6 etc.) marks the end of the somites that make neck-type vertebrae and the start of those that make thorax-type vertebrae. The HOX genes therefore correlate with the *type* of vertebra rather than simply with the number, Similarly, the beginning of expression of HOX10-type genes marks the end of the somites that make thorax-type vertebrae and the beginning of those that make lumbar-type vertebrae, and HOX11-type genes are associated with the transition between lumbar and sacral vertebrae. All of this suggests strongly that the type of HOX genes might specify what kind of vertebrae the somites will make.

This idea has been tested experimentally by genetically engineering mouse embryos so that they are missing one or more of their Hox genes and seeing what the lack of these genes means for development. In the neck of a normal mouse, there should be one first vertebra (the 'atlas') and one second vertebra (the 'axis'), each of which is specialized in a different way to allow the head to nod and rotate. After these, there are the third, fourth, and fifth vertebrae that all have a similar generic neck-vertebra shape. When the Hoxa4 gene is deleted, what should have been the third, and therefore unspecialized, vertebra develops as a second axis-type vertebra, as if its cells 'think' they were closer to the head end of the embryo than they really are. Similarly, in mice lacking both Hoxa7 and Hoxb7, what should be the first thoracic, rib-bearing vertebra develops as a neck-type, non-rib-bearing, vertebra. Again, without these genes the cells behave as if they are closer to the head than they really are. On the other hand, when Hoxa5 and Hoxa6 are removed, cells in that same region of the body 'think' that they must be located further from the head, where these genes would naturally have been turned off. As a result, what should be the seventh neck vertebra bears ribs as a first thoracic vertebra should. The same kind of thing happens lower down the body if genes towards the right-hand side of Figure 28 are removed.

In both of these segmentation systems—the clock and gradient system that divides the body into somites and the HOX gene system that tells the cells of those somites what to become—a body-scale pattern is produced by very simple local processes. In both cases, these processes use the passage of time to create a pattern in space—another example of how the complex cellular systems of developing embryos can make use of very simple principles of mathematics to create something much larger than themselves.

PART II

ADDING DETAILS

7

FATEFUL CONVERSATIONS

In principio erat verbum... *   St Jerome

The processes described in Chapters 2 to 6 have achieved a great deal from a featureless cluster of cells. The embryo has a defined long axis with a head and tail end; it has a neural tube running along the length of its back, and a gut running along the length of its front; it has blocks of tissue each side of the neural tube and an overall covering of ectoderm. What is more, cells in each level of the embryo are already constrained, through their HOX code, to behave appropriately for that level. The embryo at this stage is, however, to the adult as a painter's first sketch is to a final work: the basic structures are recognizably present but all the details have yet to emerge.

When somites first form, the total number of different cell types present in the body is still rather low. There are ectoderm cells, neural tube cells, gut cells, somite cells, and a few others, but the total is very much lower than the hundreds of different tissues to be found in an adult. The populations of cells in each of the primitive tissues of the embryo have to divide themselves up into groups that will go on to become specialized in different ways, to make different structures such as bones, tendons, muscles, and blood vessels. What is more, they have to do this in an organized way. When the embryo first faced the problem of turning a population of identical cells into two different types, way back when it divided itself into inner cell mass and trophectoderm (Chapter 3),

---

* 'In the beginning was the word...' (from Jerome's *Vulgate* Latin translation of the Bible).

it used a natural asymmetry in the cells' environment: cells with a free edge became trophectoderm. In organizing themselves into different specialized groups, the cells of the neural tube and somite also use cues from an asymmetrical environment. By this stage, though, most of the information involved comes, not from the geometrical properties such as a free surface, but from signalling molecules released by the other tissues. Using these molecules, adjacent tissues engage in a remarkable conversation that allows cells to organize one another into many different types, all precisely arranged.

An excellent example of the way that these conversations are used is provided by the specification of distinct zones within the neural tube, each destined to produce cells that serve a different purpose in the function of the nervous system.[1] The neural tube, formation of which was described in Chapter 5, runs along the midline of the embryo, between the ectoderm (primitive skin) and the rod-like notochord. It therefore experiences a natural asymmetry, with ectoderm next to its dorsal surface[†] and notochord next to its ventral surface. This asymmetrical proximity to the notochord is important, because cells of the notochord secrete a protein called Sonic Hedgehog.[‡] Sonic Hedgehog spreads out from the notochord to create a gradient of concentration, highest right next to the notochord and lower further away.[2] Cells of the neural tube are sensitive to Sonic Hedgehog, and enough of the molecule reaches the nearest to stimulate them strongly (Figure 30). As a result, these cells begin to make a new set of proteins and to become distinct from the rest of the neural tube. From this time, they are called floor plate cells ('floor plate' because the part of the neural tube nearest the notochord is its bottom surface in any animal that crawls, flies, or swims parallel to the ground. Terms like this can be confusing when applied to standing humans and it helps to picture a crawling baby, rather than a walking adult, when reading them).

[†] Reminder: dorsal = towards the back, ventral = towards the belly.
[‡] Geneticists are notorious for applying a geeky sense of humour to naming genes. The *sonic hedgehog* story began with the naming of a fruitfly mutant *hedgehog* because it caused the larva, which should have produced just a few bristles in defined stripes on the body, to make them everywhere so that it looked like a hedgehog. When it was clear that vertebrates had several genes related to fruitflies' one *hedgehog*, they were given names of species of hedgehog, at first sensible examples like *indian hedgehog* and finally, from the video game, *sonic hedgehog*. Other examples of fruitfly mutants are *tinman* (which has no heart: think Wizard of Oz), *cheap date* (which is abnormally sensitive to alcohol), and, most groan-inducing of all, *Hamlet* (in which the cell in a position meant to be occupied by Cell IIB behaved wrongly, leading the researchers to ask if the cell really was 'IIB or not IIB').

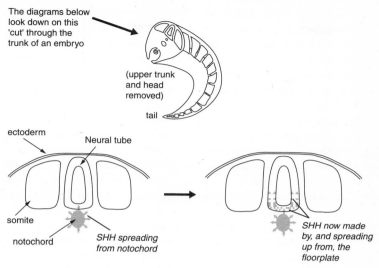

FIGURE 30 The notochord produces Sonic Hedgehog (abbreviated to 'SHH' in the labels), which spreads to the adjacent tissues of the neural tube. This causes these tissues to specialize to become floor plate, and to make Sonic Hedgehog of their own.

The importance of the notochord to the position of the floor plate is emphasized by two complementary experiments in which researchers altered the normal relationship between neural tube and notochord.[3] In one experiment, an extra notochord was surgically grafted into a chicken embryo but was put to the side of the neural tube rather than being in its normal position. In that embryo, the neural tube formed two floor plates, one at the normal site just above the natural notochord, and one along the side of the neural tube just next to the graft. This is exactly what would be expected if the notochord controls floor plate position. In the other experiment, the notochord was removed completely. Under these conditions, the spinal cord failed to form any floor plate at all.

Among the proteins synthesized by the floor plate cells is Sonic Hedgehog itself, so that the floorplate becomes a new centre for production of the molecule, which starts to spread dorsally. Sonic Hedgehog protein does not last long, so not much manages to spread far away and the concentration falls away steeply with distance. The response of cells in the rest of the neural tube depends on how much Sonic Hedgehog they receive, which will of course bear a relationship to their distance above the floor plate.

The response of neural tube cells to the concentration gradient of Sonic Hedgehog is to commit to making (eventually) one type of a few basic types of nerve cell. Examples of these include motor neurons, which send signals directly to muscles, and interneurons, which collect signals from other neurons, process them, and pass them on. Motor neurons and interneurons exist in different zones along the ventral–dorsal axis in the spinal cord.[4] The neuronal types are completely distinct: to function properly, a cell must become either a motor neuron or an interneuron, not some half-hearted hybrid that blends the characters of these cells. This raises a problem: the Sonic Hedgehog concentration gradient will be smooth and gradual, as dictated by the laws of physical chemistry, but the eventual response of the cells needs to be in sharp 'steps', one cell type then another. Translation of the smooth gradient of signal to the stepped response is performed by series of interacting genes and proteins.

The main effect that Sonic Hedgehog has is on the activation of specific genes. Different genes have different sensitivities to Sonic Hedgehog. In the developing neural tube, there is a keen gene that is activated even by low concentrations of Sonic Hedgehog, and one that is reluctant to respond and requires high concentrations.[§] The keen gene is so sensitive that it switches on initially throughout the ventral half of the neural tube. The reluctant gene is made only by cells in the ventral-most quarter of the neural tube, where the concentration of Sonic Hedgehog is sufficiently high. Importantly, the presence in any cell of the protein made from this reluctant gene switches the keen gene off in that cell. There is therefore a zone of cells making the reluctant gene immediately above the floor-plate, topped by a stripe of cells making the keen gene, with no overlap.[5] The result is the formation of distinct stripes (Figure 31).

The floor plate joins in with the notochord in patterning mainly the lower (ventral) half of the neural tube. The upper (dorsal) half also has to be patterned, and again it takes its initial cue from an adjacent tissue.[6] The nearest other tissue to the dorsal part of the neural tube is the ectoderm, from which the tube itself only recently separated (Chapter 5). This secretes another signalling protein that spreads easily into the adjacent neural tube. Here it has two main effects. The first is that it overrides any Sonic Hedgehog signals that may have spread this far. The second is that it causes the dorsal part of the neural tube to become

---

[§] These genes are Olig2 and Nkx2.2.

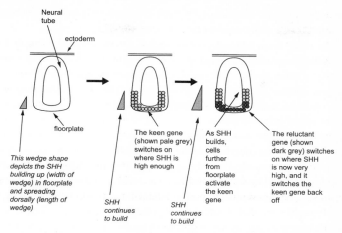

Neural
tube

ectoderm

floorplate

*This wedge shape
depicts the SHH
building up (width of
wedge) in floorplate
and spreading
dorsally (length of
wedge)*

*SHH
continues
to build*

The keen gene
(shown pale grey)
switches on
where SHH is
high enough

*SHH
continues
to build*

As SHH
builds,
cells
further
from
floorplate
activate
the keen
gene

The reluctant
gene (shown
dark grey) switches
on where SHH
is now very
high, and it
switches the
keen gene back
off

FIGURE 31 A concentration gradient of Sonic Hedgehog ('SHH' in the diagram) builds up as
the molecule is synthesized at the floor plate. It causes cells nearby to organize themselves into
distinct zones.

a signalling centre in its own right. This is very similar to what happened at the
floorplate, when the notochord made the neural tube become a signalling centre. This time, however, things are not quite as simple. At the floorplate, the
neural tube cells produced the same molecule, Sonic Hedgehog, that they had
received from the notochord. In contrast, the cells on the dorsal side of the
neural tube respond to the ectodermally derived signal, not by making more of
the same, but by making new signal proteins, WNT and BMP.[7]

WNT and BMP proteins again spread from the cells that make them to make
a concentration gradient. This gradient is used to divide the dorsal part of the
neural tube into zones in a way conceptually similar to the way that the Sonic
Hedgehog gradient was used in the ventral part (the molecular details are different but the basic idea is the same).

The neural tube, then, uses cues coming in from the asymmetrically arranged
tissues above and below it, and possibly from the somites too, to divide an initially similar group of cells into a finely patterned arrangement of zones, each
of which will make a different type of nervous tissue as the spinal cord matures.
The cells of the neural tube are not, however, just passive receivers of signals
from other tissues: they generate signals of their own. These neural-tube-
derived signals are used in the reverse direction, to pattern the adjacent tissues.

Thus the signalling is not a one-way traffic of orders, but a true conversation of statements and replies, all in the language of protein biochemistry.

When they first form, the somites that flank the neural tube are simple structures composed of a single type of cell (Chapter 6). They will, however, be the source of many of the structures in the trunk of the body, including bones, muscles, tendons, and the inner part of the skin. Like the neural tube, the somites therefore have to create internal differences and they too use signals coming from their various neighbours.

The parts of the somites that are closest to the dorsal part of the neural tube are bathed in high levels of WNT proteins, made in the dorsal neural tube (Figure 32). Somite cells already express a quite different pattern of genes to the pattern expressed in the neural tube: this is why they are somite cells and not neural cells. A consequence of this is that, when they detect WNT signals, they respond to them in a manner quite different from the way in which neural tube cells respond. The idea that one signal can have different meanings depending on the internal state of the recipient is commonplace. The sentence 'Picture me wearing suspenders' will, for example, result in a listener forming a quite different mental image depending on whether he is from America, where 'suspenders' are devices for holding up men's trousers (called 'braces' in the UK), or he is from the UK, where 'suspenders' are a device for holding up women's stockings (called a 'garter belt' in the US).** Similarly, in a committee, the phrase 'I propose we table this motion' would in the UK be understood to mean 'I propose we now discuss this', and in America to mean the exact opposite; 'I propose we lay this aside'. The large number of everyday words that can cause acute embarrassment to a transatlantic visitor, in either direction, emphasizes that, even in natural language, meaning is determined by the internal state of the receiver and is not intrinsic to the message. In biology, too, the same signal, for example WNT, can have a quite different meaning (/effect) depending on the identity of the cell that receives it.††

When somite cells receive WNT signals from the dorsal part of the neural tube, they respond by producing the proteins needed to make muscle. The

---

** A male American friend of mine once had a truly bizarre experience in a famous Edinburgh department store through not knowing this.

†† The study of the relationship between signifier and signified in human languages is semiotics: application of these ideas to biological signals and meanings is usually called biosemiotics.

range of spread of WNT is quite modest, and for this reason only the part of the somite close to the neural tube has enough signal from that source to activate muscle development. Another source of WNT causes a second centre of muscle development to form at the outer, bottom edge of the somite (Figure 32). As development proceeds, these two zones will contribute to different muscles, the zone near the neural tube giving rise to the muscles of the back and the other zone giving rise to the muscles of the body wall.[8] The zone between them, not near either source of WNT, will become the connective tissue of the skin (the dermis).

With only the signals mentioned so far, the somite would be divided into a sandwich of three zones, a skin-making one between two muscle-making ones. The somite is, however, also sensitive to signals coming from the notochord and the floorplate of the neural tube. In the parts of the somite closest to the ventral neural tube and the notochord, these signalling proteins are at levels high enough to override other signals that may be encouraging the formation of muscle and skin. They cause cells to commit to being connective tissue and bone instead[9, 10] (Figure 32).

Putting the contents of this chapter together, we see that the tissues that surround the neural tube send signals that pattern it and, as part of their response, the neural tube cells send back signals that impose pattern on the surrounding tissues. The creation of detailed pattern in a previously monotonous embryo is therefore the result of a multitude of cellular conversations.

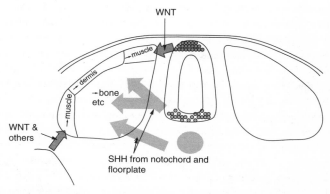

FIGURE 32 The somite is patterned by signals coming from its neighbours.

There is nothing particularly special about the tissues described in this chapter, and almost any group of closely spaced tissues could have been used to tell essentially the same story. All over the embryo, adjacent blocks of tissue use signals coming from each other to divide themselves up into different cell types, based on whether a cell is near to or far from the source of the signal. An immediate result of this process is the formation of new boundaries between the just-created cell types within the original block of tissue. If these cell types secrete different signalling proteins, the same trick can be used again to create ever more zones of different cell types. It is a powerful invention, and it is perhaps no surprise that around a fifth of the set of all human genes are devoted to producing proteins involved in some way with cell signalling.

As well as allowing tissues to be subdivided almost without limit, the use of cell communication is an excellent mechanism for coping with errors in development. Imagine an embryo in which blocks of tissue subdivided themselves in some other way that was independent of the positions of surrounding tissues—perhaps, for example, an embryo in which cells really did read instructions from some blueprint. If there were even the slightest inaccuracy in the positioning of each cell, the errors would accumulate over time and growth so that, where tissues had to come together, cell types that should be next to each other would not be, and development would fail. It may be possible, in principle, to build a very small and simple body that way, a body that involves so few divisions into different tissues and so few critical connections that errors could not mount up. Building a body with hundreds of different cell types that have to interact with each other precisely would, though, be out of the question. In a system in which tissues rely on each other for signals, on the other hand, the formation of special cell types in each tissue is positioned automatically so that cells specialize appropriately close to the tissue whose signals are used to drive specialization, even if that tissue is itself a little out of position. The organization of the system therefore keeps adapting itself to circumstances, and errors do not accumulate but are instead corrected at each stage. Therefore, unless errors in development are truly enormous, the embryo can deal with them by using the constant conversation between its cells to regulate development according to how things really are, not how they 'should' be.

The use of conversations, conducted in the language of proteins, to organize the subdivision of tissues has an interesting consequence for animal development. The range over which a protein will spread to give useable concentrations is fixed by the laws of biophysics and biochemistry, and for most proteins it is around a twentieth of a millimetre (= 50 microns). This means that a group of cells that is using this method to acquire a pattern tends to be around a twentieth of a millimetre long at the time that it acquires the pattern, whether we are talking of the spacing between dorsal and ventral parts of the neural tube, or between developing tooth roots, or developing hairs, or anything else. This has two consequences. One consequence is that not all of the embryo can acquire pattern at once. First the 'coarse-scale' patterning has to be done while the embryo is quite small then, as the parts that acquired their basic identity in this phase grow larger as the whole embryo grows larger, they can be subdivided, and so on. This is one reason that human development cannot proceed by making a minuscule but complete baby in the first week and then just growing it: successive phases of patterning have to be separated by growth. First, the embryo is divided crudely into head and trunk then, when the head has grown, it is in turn patterned to specify one region as jaw (to take an example). Then, when the jaw has grown, it is in turn patterned to specify the positions of teeth, and so on.

The second consequence is that, for any particular patterning event, such as specification of zones of the neural tube, an embryo has to be within a fairly restricted range of sizes. Thus, at this stage of this development, the embryos of a shrew, a human, and a blue whale are almost exactly the same size. The family resemblances between related animals, such as horses and whales and bats, are much easier to see when one views embryos in the womb than when one sees the adult animals galloping, swimming, and flying past.

8

INNER JOURNEYS

*Si le chemin est beau, nous ne devons pas demander où elle conduit.**
Anatole France

Most of your face came from the back of your head. Your sensory nerves and all of the pigment cells in your skin came from behind your spine. The cells making your sperm or eggs came from outside your body altogether. These facts of human anatomy, and a great many others like them, underline how much our development depends on the ability of cells to move from one place to another in an early embryo. Migrations at a cellular scale are every bit as astonishing as the great odysseys of birds and fish that have so long fascinated zoologists. Even more so given that, unlike birds, cells have to navigate accurately through an environment whose shape is changing all the time and they have to do so without the benefits of eyes, brains, or an opportunity to learn from their parents.

The problem of understanding how cells move accurately from one place to another can be split into two smaller problems; understanding how cells move at all, then understanding how this movement can be orientated in the right direction.

The ability of cells to move is very ancient and pre-dates the coming of multi-cellular animals by hundreds of millions of years. When animal-type cells first evolved to crawl over the mud of the young Earth in pursuit of their bacterial prey, they laid a foundation for movement that has remained largely unchanged

---

* If the path be beautiful, let us not ask where it leads.

to this day. The motor that drives crawling cells forward consists of several linked mechanisms. One pushes the front of the cell forwards from the inside, one makes new connections to the underlying surface, and a third uses these connections to pull the rest of the cell along so that it is not left behind by its advancing front. These mechanisms are worth exploring in some detail, because they provide a remarkable example of the ability of simple components to organize themselves into complex, higher-level systems.

The machine that pushes the front of the cell forwards from the inside is based on a fine network of protein microfilaments, exactly the same entities that were described in Chapter 5, in the context of carrying mechanical forces through sheets of cells. Microfilaments are composed mostly of a protein called actin. Individual actin molecules are small and compact but they can bind each other to make long, fine chain-like filaments (Chapter 1). New actin molecules add themselves efficiently to exposed ends of existing chains, but individual molecules come together to begin a brand new filament, only very inefficiently. This means that, in a cell, new filaments hardly ever form without some kind of help, even though any filaments that already exist tend to become longer quickly. The formation of new filaments is assisted by filament-nucleating proteins and the cell has several types, each used for a specific purpose. The filament-nucleating proteins that are important in cell crawling have the property that they work only when they can bind to existing filaments. This has the consequence that new filaments form mainly as side-branches from existing ones and, as this keeps happening, new branches eventually acquire branches of their own[1] (Figure 33).

If the filament-nucleating proteins were always active, the whole cell would become hopelessly and uselessly choked up with a tangle of microfilaments. This scenario does not occur because filament-nucleating proteins are normally inactive and have to be activated by other molecules. Some of these are tethered to the inner face of the membrane at the surface of the cell, so the filament-nucleating proteins are activated only just inside that membrane. Critically, they remain active for only a short while before 'decaying' into their inactive state. This means that filament-nucleating proteins are active mainly at the inner surface of the cell membrane and that they have time to wander only a short distance towards the centre of the cell before they are quiet again. New actin filaments are therefore begun only in a zone immediately behind the membrane at the edge of the cell.

Actin filaments that grow towards the membrane of the cell eventually collide with it and push it outwards; it is by the combined pushing of thousands of

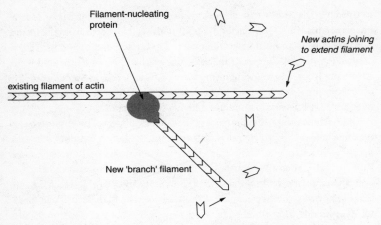

FIGURE 33 Nucleation of new actin filaments on the side of existing ones, by filament-nucleating protein complexes.

individual actin filaments that the edge of the cell advances.[2] The filament-nucleating proteins do not, however, know in which direction the front of the cell should lie. They are, after all, just large molecules and can contain no map. Many of them will nucleate branches that point the wrong way, back towards the bulk of the cell. Allowing such filaments to grow where they cannot be useful would be a waste, so the system has a way of preventing this retrograde growth. The bulk of the cell is full of 'capping' proteins that can cap the end of a filament and block its growth. Unlike filament-nucleating proteins, which are normally inactive and become activated by a process that involves molecules at the membrane, capping proteins are normally active but become inactivated by molecules associated with the membrane. Filaments growing towards the membrane therefore encounter capping protein that has been rendered inactive, and they are free to grow, but those growing in the wrong direction, back towards the cell, are quickly capped. This system therefore uses a simple asymmetry in its environment—the membrane being at one place and not the other—to orientate filament growth towards the membrane[3] (Figure 34).

The ultimate effect of the concerted action of the proteins at the leading edge is to push the membrane ever forwards by pressure from the ends of growing actin filaments. By Newton's Third Law of motion ('action and reaction are equal and opposite'), if the filaments push on the membrane, the membrane

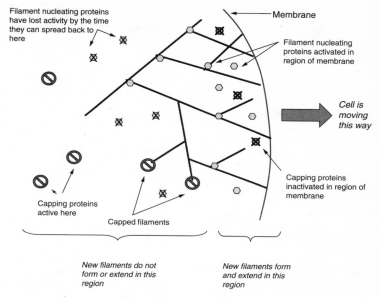

FIGURE 34 The activation of filament-nucleating proteins and the inactivation of capping proteins at the membrane create an asymmetrical environment that ensures that most new microfilaments grow towards the membrane and any growing the wrong way are quickly capped.

pushes back at the filaments. If the entire filament system were simply floating in the cell, its attempt to push the membrane forwards would simply result in the filament system pushing itself backwards. To prevent this happening, its rear end needs to be fixed, relative to the surface on which the cell is moving. This fixation is achieved by complexes of adhesion proteins that can attach the cell to the surface on which it is crawls.[4]

As well as providing the microfilaments at the leading edge with something against which to push, the adhesion complexes are important in allowing the bulk of the cell to follow the leading edge rather than being left behind. Away from the branched network found at the leading edge, actin filaments tend to form 'cables' of many parallel filaments cross-linked by a protein called myosin, which can 'pull' on the filaments and make the cables tense. These cables, similar to the cables that run between cell–cell junctions (Chapter 5), are prevented from forming within the leading edge itself, because myosin is

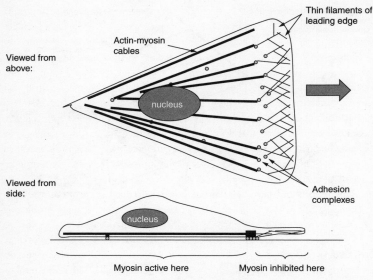

FIGURE 35 The layout of thin microfilaments in the leading edge of a cell, cell-surface adhesion complexes at the rear of that leading edge, and myosin cables leading back.

inactivated in that environment. Further into the cell, though, myosin is free to act and to organize actin into cables (Figure 35). The cables, therefore, run from the rear of the leading edge, the most forward part in which they are able to form, backwards throughout the whole cell. They also bind the adhesion complexes so, as myosin causes the cables to become tense and to pull, the cables tend to pull the rest of the cell forwards towards their front ends. Shortening of the cables by simple disassembly also helps.[5] The adhesion complexes therefore carry two sets of forces: the network of leading edge actin filaments pushes off them to advance the front of the cell, and the thick cables running to the rest of the cell pull on them, to draw the rest of the cell forwards. These forces have actually been measured, by allowing cells to grow on very thin and elastic surfaces that become visibly deformed by all the pushing and pulling.[6] The adhesion complexes are strongest just behind the leading edge and become weaker as the cell moves over them and they become more and more distant from the leading edge; the weakened adhesions further back let go and the cell as a whole can move forward.

In summary, then, a relatively simple set of components comprising actin and a few proteins that regulate actin's behaviour is capable of using a simple asymmetry, to do with how close a given part of the cell is to the membrane, to organize itself into a machine capable of moving a cell. Each of the proteins is, of course, encoded by a gene in the nucleus of the cell but, beyond doing the job of ensuring that these proteins are produced, the nucleus and its genes play no role in actually organizing the machine. This point has been demonstrated beautifully in experiments on fish cells, in which small fragments of cell that contain no nucleus, and therefore no genes, are shown to be completely capable of crawling. The 'intelligence' to do that is a property of the set of proteins. Not a property of one protein—for no one of these proteins can do anything very useful on its own—but of the set.

This, then, is an answer—in outline at least—to the first part of the problem of cell migration: how cells can move at all. The second part of the problem is how the movement of cells can be guided in a particular direction. Embryos use two broad approaches, sometimes separately but often together. One works by presenting cells with a choice between surfaces of different adhesive properties, while the other works by presenting cells with molecules that increase the activation of the filament-nucleating protein at the leading edge.

Guidance by adhesion can be understood most easily in the context of some simple culture experiments done in the closing decades of the twentieth century. In the laboratory, cells can be grown on a variety of surfaces, ranging from simple plastic to complex biological molecules. The adhesiveness of these surfaces can be placed in rank order simply by noting how hard one has to blast liquid at the cells in order to wash them off the surface. When cells are presented with a 'patterned' surface, some parts of which are more adhesive than others, they move randomly until they find themselves at a boundary. At the boundary, they have a choice of surface and they choose the more adhesive. The explanation is easy to understand. Consider a cell that is placed exactly across a boundary between a slippery and an adhesive surface (Figure 36). By definition, the part that is on the more adhesive surface will form more adhesive connections than the other part. The fine filament network of that part of the leading edge will therefore have a good anchor against which to push and the leading edge of this part of the cell will advance quickly. The leading edge on the slippery surface will not have such a good anchor, so some of its effort will be wasted in moving itself backwards. Similarly, the actin–myosin cables that pull the bulk of the cell forwards will have many good anchors on

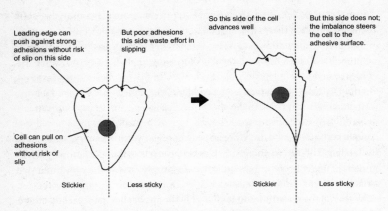

Leading edge can push against strong adhesions without risk of slip on this side

But poor adhesions this side waste effort in slipping

So this side of the cell advances well

But this side does not; the imbalance steers the cell to the adhesive surface.

Cell can pull on adhesions without risk of slip

Stickier

Less sticky

Stickier

Less sticky

FIGURE 36 A boundary between two surfaces, one of which supports adhesion well and the other of which supports it poorly, can steer a cell in the direction of enhanced adhesion.

the adhesive side of the cell but not on the other, so the bulk of the cell will move better this way. The leading edge and the bulk of the cell therefore both advance best towards the more adhesive surface and the cell is therefore guided this way.

For some years, scientists voiced their scepticism about whether the explanation for this really is all mechanical, or whether signals from adhesion complexes might be encouraging the filament-nucleating proteins and therefore guiding the cells chemically rather than mechanically. These objections were silenced by an elegant experiment[7] in which researchers placed a tiny glass bead, coated with an adhesive molecule, on to the *top* surface of a leading edge. The bead stuck to the adhesion complexes of the cell and the leading edge network tried to use it as an anchor against which to push. The bead was not attached to anything else, so this push simply moved the bead backwards and did nothing to help the leading edge advance. The direction of cell movement was therefore unaffected. If the experimenters stopped the backwards movement of the bead by physically restraining it with a fine glass thread, the leading edge near it advanced forwards quickly and the cell steered towards that side. In both cases, the bead was bound and any chemical signals would therefore have been the same, but only when the bead was restrained were mechanical forces present. This showed, very effectively, that mechanical forces really can steer cells in the absence of any chemical differences.

Even though mechanics can account completely for guidance in some cases, chemical signalling is very important in others. Some molecules on the surface are not used for adhesion, but are instead recognized by receptors in the cell membrane that activate signalling pathways inside the cell. Some receptors and pathways act to stimulate, locally, the activity of filament-nucleating proteins and therefore the local advance of the membrane. Others act in the exact opposite manner and inhibit the pushing forward of a leading edge; these tend to repel cells from that type of surface.

Experiments with beads and patterned plastics are all very well, but how is this tendency of cells to choose stickier surfaces relevant to the complex environment of an embryo? In particular, if cells simply choose the more adhesive surface, how can different cells take different routes, which might even cross each other, at the same time? At least part of the answer lies in the fact that there are many different types of adhesive complex, each of which adheres most strongly to a specific surface protein. For example, the adhesion complex α6β1-Integrin binds to a surface protein called Laminin, whereas the similar complex α5β1-Integrin binds instead to a quite different surface protein called Fibronectin. Different types of cells carry different combinations of adhesive complex. One carrying the α6β1-Integrin would perceive Laminin as a strongly adhesive surface, whereas a cell carrying only α5β1-Integrin would not. This means that different cell types will make different decisions about direction, even when they are crossing the same set of surfaces.

The stationary cells of the developing tissues of an embryo synthesize surface molecules such as Laminin and Fibronectin. Different types of cells secrete different combinations of molecules. This means that the different tissues of a developing animal appear, to a crawling cell, as a richly patterned landscape of more or less adhesive surfaces, some of which might also have chemical signalling molecules on them that are either attractive or repulsive. Different types of crawling cell will see this landscape in different ways, according to the adhesion complexes and receptors they carry, and what might be a highly attractive surface for one cell type may completely repulsive to another. It is another example of the meaning of a sign depending on the interpreter: our identifying the signifier is not enough, on its own, to indicate what is signified. In this way, different types of cell can follow different routes through the embryo.

Much is still being discovered about the pathways that lead migrating cells through the embryo, but enough is already known to illustrate the general principles outlined above with some real examples. Each of the cases described here concerns a population of cells that detaches from the most dorsal part of the neural tube and migrates, splitting into streams of loosely connected cells, to reach many different locations in the body. These cells are known collectively by a name based on their point of origin, the 'neural crest'. The neural crest cells in the main trunk of the body have four main fates: they make the sensory nervous system, which brings in sensations of heat, pain, touch, and position to the spinal cord and brain; they make the autonomic nervous system, which is responsible for running many functions of our organs that are outside our conscious control, such as rate of heartbeat; they make the core of the adrenal gland, which releases hormones such as adrenalin; and they make the pigment cells that protect the skin from damage by ultraviolet light. These very different fates are associated with different routes of migration.

The first neural crest cells to detach themselves from the neural tube make receptors for the attractive surface proteins Laminin and Fibronectin (Figure 37). They therefore choose migration routes that contain these proteins. The neural crest cells also make receptors for a molecule called Ephrin (Figure 37), which is repulsive to them.[8] The somites that flank the neural tube make plenty of attractive Laminin and Fibronectin, as does the bottom of the ectoderm—the primitive 'skin' that lies above the somites and neural tube (Figure 37). Many somite cells also express Ephrin, and are therefore repulsive to the early-

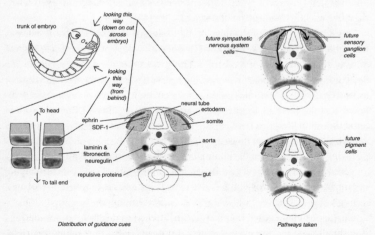

FIGURE 37 Neural crest cells migrate along paths that are defined by molecules secreted by the tissues through which they travel.

emerging neural crest. In the posterior (tailward) half of the somite, which will form the bony part of a vertebra, Ephrin is expressed by all cells, many of which also express additional repulsive molecules.[9] In the anterior (headward) half of the somite, the expression of Ephrin is more complicated: the part that lies just under the primitive skin of the back makes Ephrin, but the rest of the somite does not.

When a neural crest cell emerges from the neural tube next to the anterior half of a somite, it detects Ephrin to the left of it, Ephrin to the right of it, Ephrin behind it (in the posterior half of that somite), and Ephrin ahead of it (in the posterior of the next somite headwards). It is therefore repelled from migrating in any of these directions. The cell is already right at the dorsal edge of the embryo, so it cannot migrate any further that way. The only way open to it is therefore to dive down ventrally, into the main bulk of the somite where it can find plenty of attractive Laminin and Fibronectin and is free from assault by Ephrin. Thus the neural crest cells are guided to enter the loose-packed tissue of the anterior half of each somite, where they can move around freely.

Beyond the somite, deeper into the body of the embryo, there are tissues that secrete a protein called Neuregulin (Figure 37). Some of the neural crest cells that enter the somite make receptors for neuregulin and are attracted by it. These cells therefore move right through the attractive cells of the somite to enter the *even more* attractive tissues[10] that lie deeper in, next to the aorta, the main blood vessel that runs along the length of the trunk (Chapter 9). Deeper still are the tissues that surround the developing gut. These make molecules that are strongly repulsive and therefore ensure that cells streaming down from the somite in the tissues around the aorta do not overshoot. Trapped by the attractiveness of their present location and the repulsiveness of the deeper tissues around the gut,[11] the neural crest cells remain where they are, in the tissues next to the aorta. The aorta secretes a protein that causes neural crest cells to become the sympathetic nervous system, which controls the function of our internal organs without our having to think about them.

Having the maturation of the migrating neural crest cells to nervous tissue triggered by proteins that are produced by tissues at their final destination is an elegant method for error control. If the final settling and maturation of migrating cells were instead determined by something intrinsic to them, for example a 'clock' that caused them to mature after so many hours of wandering, then they would mature even if they had not succeeded in reaching the right place, and the body would be full of odd islands of tissue produced by cells that were lost but decided to make a tissue anyway. Obviously, this would not be a good thing. Making maturation depend on signals from the correct destination ensures that cells that have not found their destination continue to seek it rather than give up and mature in the wrong place.

While the earliest neural crest cells to leave the neural tube tend to go on through the somites to the region of the aorta, ones that leave a little later tend to stop in the somite. As they leave the neural tube, these cells express a receptor for a protein called SDF-1 (Figure 37), which is made by the cells of the somite. When these neural crest cells enter the somite, the SDF-1 they find there causes them to stop migrating and to settle in the tissues of the somite itself.[12] Here, they aggregate together and eventually make a small lump of the nervous system—a sensory ganglion—which is concerned with relaying sensory information into the spinal cord.

Once the migrations described above have been going on for a while, a third population of neural crest cells emerges from the neural tube. These are different again. They still make the proteins needed for cell movement; they still make receptors for Laminin and Fibronectin and they still make receptors for Ephrin. Unlike the early-emerging cells, though, they make a set of internal proteins that cause them to interpret Ephrin as attractive.[13] When they emerge from the neural tube, they see Ephrin to the left of them and Ephrin to the right of them, in the parts of the somite just under the ectoderm of the back, so they immediately launch themselves into these tissues. They therefore follow a quite different route from the early-emerging cells, and stay just under the embryo's ectoderm. This is important, because these late-emerging cells are already committed to becoming the pigment cells of the skin.

The prior commitment of late-emerging neural crest cells to be pigment cells raises an important point. In principle, there are two possible ways in which the population of neural crest cells could follow different pathways to become different tissues. By analogy with old debates on human behaviour, the two possibilities can be labelled 'nature' and 'nurture'. In the nurture method, once assumed to be true, cells that were initially identical would take pathways at random and they would decide what kind of cell to be, according to the environment in which they ended up.[14] In the nature method, initial differences in the gene expression patterns ('natures') of the cells would cause the cells to follow the correct pathway to find the place that suited their already-chosen fates. It is clear from the way that cells express different sets of proteins that cause them to interpret guidance cues differently, that the 'nature' model is essentially correct[15] although, as noted above, the final environment is still needed to signal cells to stop migrating and get on with the business of realizing the fate they had already chosen. The discovery that the 'nature' explanation is the correct one does not solve the problem of how the neural crest cells decide their future fates, but just puts the problem back to the time before they leave the neural tube.

The preceding pages have concentrated on only the neural crest cells of the main trunk of the body. Neural crest cells that arise in the neck, head, and tail regions have a different range of fates available, and these include making the nervous system of the gut, bony structures of the face, pigment cells of the iris of the eye, parts of the ear, and parts of the heart. Again, these cells follow different routes according to what they will make. It would become tedious to recount all the migratory pathways and cues here, but each is about as complicated as the trunk pathways that have been described in detail. This complexity means that correct navigation depends on many components being made by the right cells at the right time. Genetic mutations that prevent this from happening, for example by damaging a gene coding for one of the proteins involved, cause specific congenital defects of human development. These defects are known collectively as neurocristopathies (a word derived from 'neural crest' and '-path', indicating disease). One set of mutations, for example, disrupts the signals that should bring neural crest cells from the neck and tail regions into the gut so that they can make the gut's nervous system.[16] In affected individuals, part or all of that nervous system fails to form. That means that food and waste are not moved properly along the gut, giving rise to a very severe constipation called Hirschsprung disease: most cases have to be treated surgically, early in childhood. Yet another set of mutations, in genes that control the internal state of neural crest cells and their maturation, causes Waardenburg syndrome.[17, 18, 19, 20] In this disorder, defects in neural crest migration and/or maturation, especially of pigment cells, cause defects in hearing, pigmentation of the iris, pigmentation of hair (a white forelock is particularly common) and, for some mutations, Hirschsprung's disease as well. The neural crest cells of the head require certain proteins, such as Treacle, to produce enough new cellular material to keep up with their multiplication as they prepare to make the structures of the face. Mutations of the gene encoding Treacle results in many cells becoming stressed by running short of essential components and dying. Too few survive to make a normal face, and the result is the unusually shaped face of a person with Treacher-Collins syndrome,[21] with slanting eyes, under-developed cheeks, a small lower jaw, drooping eyelids, and small or absent earlobes.

The nature of the genes that are mutated in these neurocristopathies illustrates an important point about the link between development, genetics, and disease. The popular press, and even some poorly-written technical articles, often use a phrase like 'the gene for such-and-such a disease'. Where the disease involves the loss or malformation of some structure, such as the face, the writers tend to imply that the normal function of the (unmutated) gene is to make that structure. Yet when we actually examine the proteins made by the genes whose mutation causes neurocristopathies, we find that they form parts of complex, multi-component mechanisms that perform tasks such as cell guidance. These tasks are at a much lower level, and take place at a much smaller scale, than the development of a large structure such as the face. The function of the protein Treacle, for example, is not by any stretch of the imagination to make a face: in fact, Treacle plays a role in ensuring the efficient production of ribosomes, the molecular machines that translate messenger RNAs into proteins (Chapter 1). This is a simple biochemical task that has nothing directly to do with making faces. It so happens, though, that with Treacle not working properly, neural crest cells of the head run short of ribosomes quickly and become stressed—so stressed that they die.[22] This has the consequence that there are too few cells to make a normal face. The final defect leads to the illusion that the normal function of Treacle is to make a face, when all it really does is help make ribosomes.

The general point made by the Treacle example above—that it is a serious misinterpretation of evidence to conclude that the function of a gene is to make a particular body structure—applies equally to almost all genes. The misinterpretation of what genes really do has led to some fun science fiction, but also to very unrealistic expectations of how easily we might engineer novel body shapes or parts by playing with a few 'designer genes'. Body parts are made by networks of interacting proteins, each of which is encoded by a gene, and if we are to understand how the body parts are made and how we may, for good or foolish reasons, change that, we have to understand human development, not at the level of individual genes, but at the level of the interacting networks.

The neural crest has been chosen as an exemplar of cell migration, but it is by no means the only migratory cell type. Other long migrations are made by the cells that will give rise to sperm and eggs (Chapter 10), and cells that will give rise to the blood system (Chapter 9). Short-range migrations are made by a vast range of cell types as they organize themselves into tightly knit groups to make bones and parts of organs (Chapter 12). During the development of the nervous system, small parts of cells migrate to create the long, thin cable-like cell projections (axons and dendrites) that connect nerve cells to each other and to sensors and muscles (Chapter 13). Even in a grown adult, defensive cells of the immune system can migrate towards sites of infection (Chapter 17). Less helpfully, many cancer cells re-activate migration machinery and spread from the initial site of a tumour to other places in the body in a process called metastasis. Much of the intense research into the mechanisms of cell movement in normal development has been funded by cancer research charities, in the hope that by understanding normal cell migration we may learn how to stop the dangerous metastasis of cancers. This is just one example of how research that can seem quite abstract 'ivory-tower' embryology connects very firmly with urgent problems that affect real people's lives.

9

PLUMBING

Man, an ingenious assembly of portable plumbing.
Christopher Morley

Cells are very small objects, typically about a hundredth of a millimetre across. The proteins that run the reactions inside them are much tinier, around ten-millionths of a millimetre, while the molecules of water in which they are dissolved are much smaller again. When viewed at these very fine scales, the interior of a cell teems with motion. This has nothing to do with the fact that cells are alive, for it would be as great in a dead cell or even in soup at body heat: it arises from basic physics. At any temperature above absolute zero, molecules show random vibrations and movement ('temperature' is just a measure of this energy of motion, averaged over the population of molecules). As they move, molecules collide occasionally and, when they do this, they bounce off one another. If any larger molecules, such as proteins, happen to be dissolved in the water, the water molecules will collide with them too, giving some of their momentum to the larger molecules so that these too are in constant random motion.

The effect can even be seen to act on much larger and heavier objects than proteins, such as grains of smoke and pollen. Indeed, the Roman poet-scientist Lucretius suggested as early as 60 BC that the random dance of smoke grains in the air might imply rapid movement of 'atoms' (his word) that are too fine to see but that collide randomly with the smoke.[1] The effect was first observed in liquids by Jan Ingenhousz in 1785 and was again described once more, forty-two years later, by the Scottish botanist Robert Brown. The movement of objects dissolved or suspended in liquids is now named 'Brownian motion' after him (which seems a little hard on Ingenhousz), while Einstein is generally given credit for understanding that it arises from the jostling of invisible atoms/ molecules (which seems equally hard on Lucretius).

The reason why this is important to a developing embryo is that the random motion of molecules creates an automatic mechanism for transporting dissolved substances, such as food, oxygen, and raw material for construction, to the places they are needed. An enzyme that requires a particular raw material has only to wait and the random motions of the enzyme and the raw material will soon bring them into contact. This process works well over small distances, but the random motions, involving many reverses of direction, makes transport significantly less efficient as the distances become larger. This creates a problem for a growing embryo because, with the source of food and oxygen being in the walls of the uterus outside, cells deep inside the expanding body are in danger of being starved of these essential materials. The maximum distance a typical mammalian cell in a solid tissue can be from a source of raw materials, and still obtain enough by random thermal diffusion, is a few tens of cell diameters (a few specialized cells, particularly in the skeleton and its associated tissues, tolerate greater distances). An embryo that will build a body larger than that—and the centre of an adult human torso is approximately thirty thousand cell diameters from the overlying skin—has to develop a system that effectively brings nutrients deep inside the body, close enough to every cell that random thermal diffusion is able to bridge the remaining distance. In humans, as in all vertebrate animals, the problem is solved by the circulation of a carrier fluid, blood, through fine tubes that ramify almost everywhere.

A working blood system requires four main components: the liquid component of blood in which most nutrients and toxins are carried; freely-floating blood cells that are specialized for carrying oxygen (which is troublesome to transport in free solution); a closed system of tubes to bring the blood to the tissues; and a pump to keep the blood circulating between tissues that load useful substances into it and tissues that use them up. In a developing embryo, all tissues of the body are consumers of nutrients and producers of wastes. At the placenta, the embryo's blood runs very close to the mother's, which is being replenished and cleaned by her lungs, gut, liver, and kidneys. Since molecules can flow fairly freely between embryonic and maternal blood systems, their concentration tends to equalize so that the exhausted embryonic blood collects oxygen and food at the placenta and gives up its wastes there. Embryonic blood, therefore, needs to flow in a circuit that takes it from placenta to embryonic body and back again. This requires the development of a system of blood vessels, and a pump.

The vascular system of a human embryo forms in two main parts. One part is located outside the embryo's body, in the yolk sac, while the other is located fully inside the embryo. Both are important, and they connect together to make a single, integrated system but, for simplicity, this chapter will concentrate mainly on the part of the circulation that develops inside the body.

The cells that will form the first blood vessels of the embryo proper are first detectable in the mesoderm* out at the edge of the embryo at around the time that the neural tube forms[2] (Chapter 5; Figure 38). Their formation, and also their proliferation, is driven by a signalling protein, VEGF, that is secreted by other parts of the mesoderm and by the endoderm below it.[3, 4] In response, these cells, heamangioblasts,† begin to multiply and to express proteins characteristic

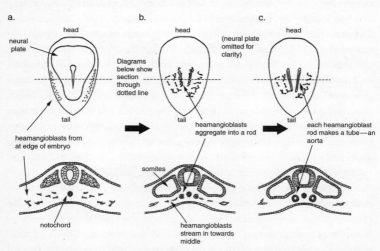

FIGURE 38 Formation of the embryo's first arteries, the aortae: (a) shows an plan view of embryo, and a section through it, during the early stages of somite formation; in this embryo, haemangioblasts form in the outer parts of the mesoderm. In (b), these cells have begun to stream in and to aggregate in rods each side of the notochord, while in (c) the rods have hollowed out to make hollow vessels, the aortae which, at this stage, remain separate and distinct from one another.

---

* Reminder: the mesoderm is the loose-packed, middle layer in the sandwich between ectoderm and endoderm, and was described in Chapter 4.
† 'heam-' connotes blood, 'angio-' connotes blood vessel, 'blast-' is a word for an embryonic cell.

of blood vessels. They also become migratory, and they can detect, and migrate towards, signalling proteins that are produced by somites. The haemangioblasts therefore migrate from their site of origin, out at the edge of the embryo, in towards the mid-line of the trunk[5, 6] (Figure 38b).

If there were no further signals, the development of the haemangioblasts would stop there. Their continued development depends, like so many things near the mid-line of the early embryo, on signalling by the Sonic Hedgehog protein made by the notochord.[7] In the presence of Sonic Hedgehog, the heamangioblasts adhere to one another to form a network of connected cells. The ones that have converged under the somites form such a dense network that it appears as a solid rod along each side of the embryo (Figure 38b). These rods go on to hollow out from a solid rod of cells into a tube (Figure 38c). The mechanism responsible for this conversion is reminiscent of the earliest days of the embryo, when trophectoderm and inner cell mass became distinct from one another (Chapter 3). Cells on the outside of the rod have a surface that is in contact with ordinary mesoderm rather than with other heamangioblasts: in response to this asymmetry, these cells survive and make a tight, cylindrical sheet. The cells on the inside of the rod are surrounded only by other heamangioblasts, and this lack of asymmetry causes them to activate a 'cell suicide' programme that eliminates them to leave a clear space (the role of cell suicide in development will be explored more fully in Chapter 14). The result of this is that what was a solid rod of cells is transformed into a hollow tube, the walls of which are polarized (Figure 39). The two tubes formed this way, one to the left of the mid-line and one to the right, are called aortae.

The importance of Sonic Hedgehog from the endoderm is underscored by two types of experiment, done by the same research team.[8] In the first, Sonic Hedgehog was removed by removing the endoderm itself, thus removing the main source of the molecule: the aortae failed to form. In the second experiment, pure Sonic Hedgehog was added back, artificially, to an embryo with no endoderm: blood vessel formation was then rescued. The importance of Sonic Hedgehog is confirmed by the observation that, even in simple culture dishes, haemangioblasts can be induced to make a network of vascular tubes just by treating them with Sonic Hedgehog protein.

Given that the haemangioblasts become mutually adhesive under the influence of signalling proteins, there might be a real risk of them coalescing to make one single, mid-line rod. This is prevented by short-range signalling proteins, such as Noggin, from the notochord. Given the limited space under the notochord,

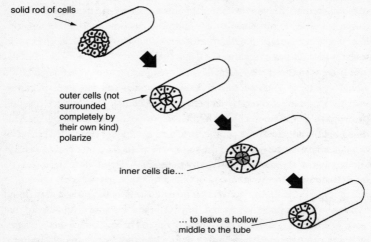

solid rod of cells

outer cells (not
surrounded
completely by
their own kind)
polarize

inner cells die...

... to leave a hollow
middle to the tube

FIGURE 39 One idea of how the solid rod of heamangioblasts hollows out to make an artery. In this idea (which has been observed in other animals but not verified in humans), the outer cells of the rod, which detect the asymmetry of having one surface free of contact with other haemangioblasts, polarize to form a leak-proof tube. The cells in the inside, completely surrounded by similar cells and unable to make contact with the tissue outside the tube, deliberately die to leave a hollow through which blood will one day flow.

between it and the endoderm, this effectively bans blood vessel formation at the mid-line and ensures that two separate aortae form, one each side of the body[9] (Figure 40). This is the correct arrangement for the early embryo, although it will change later as the foetus matures.[10]

Blood vessels of the body come in two types, arteries and veins. Arteries are relatively small-bore, thick-walled pipes that carry high-pressure blood from the heart to the small blood vessels of the tissues; veins are large-bore, thinner-walled pipes that return the blood, its pressure mainly spent, to the heart. It turns out that haemangioblasts are already committed to being either artery or vein even before they have migrated towards the centre line to make the great blood vessels, and they express subtly different sets of proteins, especially those connected with cell-to-cell interactions. At least in fish, where it has been studied most closely, the choice of whether to become an artery-type haemangioblast or a vein-type haemangioblast is controlled, yet again, by levels of proteins of the Hedgehog family. If a cell receives high concentrations of these signalling proteins, it develops early, migrates early, and takes on an arterial

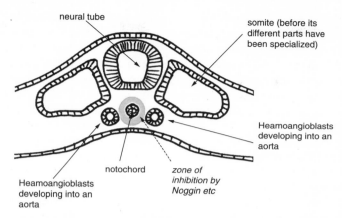

FIGURE 40 Noggin, secreted by the notochord, prevents the haemangioblast rods from coalescing along the mid-line and forces them instead to form two distinct rods, which later hollow out to form the left and right aorta, each side of the midline.

fate; if it receives only modest concentrations, it develops later, migrates later, and has a vein fate.[11] The different sets of proteins expressed by artery-building and vein-building cells allow them to recognize and stay with their own kind. Cells already set to be veins migrate along slightly different routes but make blood vessels by the same basic process. Along the length of the embryo, the two great veins form parallel to the aortae, but further from the mid-line (naturally, because the further from the midline they are, the lower the concentration of Hedgehog signals and the more 'vein-friendly' is the environment). In the tissues, the arterial and vein systems send out fine branches that connect up, forming a network of small blood vessels, the walls of which are thin enough for oxygen and nutrients to pass through.

Formation of a network of blood vessels is of little use unless there is a heart to pump blood around. The heart is therefore one of the first organs to form in the embryo, and the first to reach a usefully functional state. The heart begins to form around nineteen days after conception, about two days after the great vessels start to form. In a process similar to the way that heamangioblasts came together to form tubes on each side of the mid-line, the cells that will make the heart come together as another pair of tubes. These cardiogenic cells also come from the mesoderm in the outer edges of the embryo but, in the case of the heart, the cells are from the outer edges of the head part of the embryo rather than its trunk (Figure 41a). Their formation is restricted to this location by the interplay of three families of secreted proteins.[12] First, there are signals made by the endoderm at the very edges of the embryo; they stimulate the formation of cardiogenic cells. Second, there are the WNT proteins secreted by the neural tube (as described in Chapter 7): they act in the opposite direction, to inhibit the formation of cardiogenic cells. Third, there are proteins that are made in the endoderm in the head end of the embryo; they oppose the action of WNTs (in these cells, anyway). The head is therefore the only place that WNT is inhibited enough to allow cardiogenic cells to form.[‡] The restriction of the positive signals to the very edge of the embryo means that cardiogenic cells are made in a crescent around the outer edge of the head, and not in the middle.

Like the haemangioblasts that made arteries and veins, the cardiogenic cells become migratory and find and stick to each other. They form tubes, and the ends of these tubes that face the trunk of the body associate with the aortae, joining onto them to form continuous lines of cells. By the time this process is happening, the embryo is elongating quickly and the geometry of its ends is changing. In the closing part of Chapter 5, mention was made of the way that the ends of the embryo curl down over the endoderm, transforming it into the gut tube. The other result of this curling is that the crescent of tissue at the edge of the head, in which cardiogenic cells are located, is drawn down over the front of the embryo until it lies below the gut, at about the level of the future chest (Figure 41b, c). Here, the cardiogenic cells, which are in the act of making two separate, parallel tubes, are forced together in one place. When the parallel dorsal aortae formed dorsal to the gut, they were prevented from fusing together by inhibitory signals from the notochord. Where the heart tubes find themselves parallel, ventral to the gut, there is no notochord and there are no inhibitory signals. The tubes are therefore free to meet and they do, fusing together to make a single, large-diameter tube: the primitive heart.

The curling of the embryo over the end of the gut brings the ends of the aortae, to which the heart tubes have already joined, down with them. The end of the left aorta is drawn down around the left side of the gut, and the end of the right aorta is drawn down around the right side of the gut, so that the net effect

---

[‡] The positive signals are carried by BMPs: the rescuing of cells from the action of WNT is done by Crescent and Cerberus.

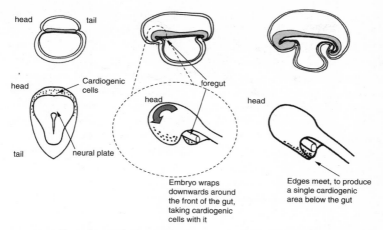

FIGURE 41 Three phases in formation of the heart area. The top row of pictures, which show the embryo from the side, have already appeared in Chapter 5; the lower row of pictures shows a top view (left) or a top-and-side view (middle and right). In the first stage (left), when the embryo is still quite flat and before much elongation takes place, cardiogenic cells are to be found in the mesoderm beyond the neural plate in the head area. Later (middle), as elongation takes place and starts to drag the ends of the endoderm into a gut tube, the embryo curls down. When its edges meet under the gut (right), they bring the cardiogenic cells together into a single area below (ventral to) the gut: this is where the heart will form.

is to surround the gut in an arterial ring (Figure 42).§ Later, the notochord stops secreting signals that prevent the two dorsal aortae that run above the gut from fusing, and these two vessels join to become one.

The early heart tube consists simply of the type of cell that lines blood vessels, with no evidence of muscles or any of the other specialized command-and-control tissues that are needed for an organ that can beat and/or pump. Within a few days of its formation, however, the heart tube attracts cells that can form heart muscle and the other tissues that a heart needs.

Heart muscle cells have the interesting property that they can twitch all by themselves, without needing to be connected to a nervous system. When they join together, they communicate electrically so that their individual twitching is synchronized. Researchers who work with embryonic stem cells often see

§ The ring is the 'aortic arch': properly, this is called the 'first aortic arch' because, later in development, a series of further aortic arches form parallel to it. The anatomy made sense when our ancestors were primitive fish and the arches served a series of gills.

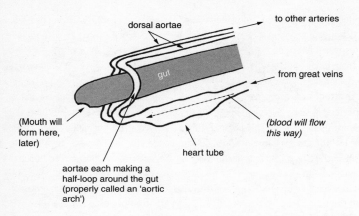

FIGURE 42 The same folding process that brings the cardiogenic cells to the bottom of the embryo drags their connection with the head end of the aortae down too, resulting in the formation of a loop of arteries around the gut, the aortic arch. In this cartoon, the embryo has been 'straightened' to make the essentials of its anatomy more clear.

this happening before their very eyes. If embryonic stem cells are cultured in conditions that allow them to re-start their development and try to make the tissues of a body, some of them become heart cells. Within a few days, the bottom of the culture dish is peppered with islands of regularly twitching cells, in a sea of other cell types.[13] It is a remarkable sight, and an endless source of fascination even to those of us lucky enough to see it happening every day.

Although a simple muscular tube is a far cry from the intricate anatomical complexity of the mature human heart, it seems to be good enough to move small volumes of blood around the early embryo. It is difficult to watch the hearts of human embryos, for obvious reasons, but the very similar heart tubes of young fish embryos are easy to study because most fish develop outside their mothers and their embryos are transparent. Also, some types of fish can be genetically engineered so that their heart tubes fluoresce bright green, allowing them to be observed easily. Filming the hearts of these embryos at high speed reveals a pumping mechanism of surprising subtlety. It used to be assumed that tubular hearts moved blood by simple peristalsis, that is by a wave of muscle contraction and tube narrowing sweeping along the tube pushing fluid ahead of it. Careful study[14] of the movies has instead revealed a more efficient process that makes use of the way that pressure waves are reflected at the end of a pipe.

Reflection of waves of fluid pressure at the ends of pipes, or at places that pipes change diameter or elasticity of walls, is a commonplace phenomenon. It is responsible, for example, for reflecting sound waves back down an organ pipe from its open end, where the part of a wave reflected downwards interacts with the part still coming up; if the wavelength is just right for them to be in step, the waves add together to make a loud sound (the wavelength that will do this perfectly depends on the length of the pipe, which is why different length pipes are used to make notes of different wavelength, or 'pitch'). Reflected waves of fluid can be powerful. This is illustrated at a human scale by a recollection in the book *Signalman's Morning*, an autobiography by the retired railwayman Adrian Vaughan:[15] in the book is a story of how an engineer testing a valve in a large water main opened and closed the valve rapidly several times. When the reflected pressure wave caused by the first opening came back and met the next wave, the pressure was enough to burst the iron water pipe and create a few hours of chaos on the main railway line between London and Bristol.

Although there are no valves or open ends in the young blood system, there *is* a sudden change of diameter and wall elasticity where the great veins join the heart. This is enough to reflect pressure waves, and the young heart makes very clever use of this fact. Its cycle of beating begins when a small zone quite near the tail end of the heart contacts, narrowing the tube locally (Figure 43). Contraction spreads from this point in *both* directions. A headward spread of contraction makes intuitive sense, because it will squeeze blood forwards, but spread of the contraction tailward seems counter-intuitive, since it seems set up to move blood the wrong way. The pressure wave this contraction creates, though, travels tailward but reflects strongly from the junction with the great veins. As the reflection happens, the contraction at the tail end of the heart relaxes, allowing the tube to open up and start to suck blood forwards. Blood is therefore both pulled forward by suction and pushed hard forward by the reflected pressure wave, which is now travelling headward, rushing into the space that is opening up behind the wave of contraction that had been travelling headward all along.

The simple heart is enough to push blood around the embryo's body itself and to push it out to the placenta and around the yolk sac. Later on in development, the basic heart tube undergoes a very complicated series of foldings, joinings, and changes of connection so that what was a simple tube that pumped blood forwards becomes a four-chambered organ with many different valves to

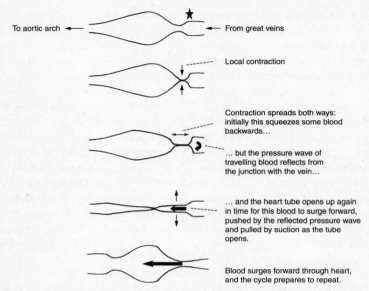

To aortic arch ◄———

★

———► From great veins

- - - - Local contraction

Contraction spreads both ways:
initially this squeezes some blood
backwards...

... but the pressure wave of
travelling blood reflects from
the junction with the vein...

... and the heart tube opens up again
in time for this blood to surge forward,
pushed by the reflected pressure wave
and pulled by suction as the tube
opens.

Blood surges forward through heart,
and the cycle prepares to repeat.

FIGURE 43 The early heart is anatomically very simple, with no valves or elaborate control systems, but nevertheless manages to pump efficiently thanks to the reflection of the tailward pressure wave at the heart–vein junction (the position of which is marked by an asterisk in the first figure).

pump blood twice, once through the lungs and then again through the rest of the body.

Once the aortae, the great veins, and the heart have been produced, the embryo begins to use two new methods for making further blood vessels. The first method sends out side-branches from existing arteries and veins; these grow into tissues and join, where they meet, making continuous tubes through which blood will flow. The aorta, for example, sends out branches that push their way between the somites and ramify through the neural tube and the muscles of the back, finally connecting to branches to the great veins. It also sends out branches to organs that develop near it, such as the kidneys (Chapter 10) and the gonads (Chapter 11). In some areas of the body, such as the neck, the arterial branches that pass between the somites send out side branches towards each other so that they all connect up. This means that what was a series of parallel arteries

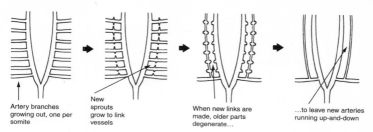

Artery branches growing out, one per somite

New sprouts grow to link vessels

When new links are made, older parts degenerate...

...to leave new arteries running up-and-down

FIGURE 44 An example of the large-scale remodelling of blood vessels that takes place during human development. In the neck region, as in the trunk, a series of arteries buds out between body segments. These, in turn, send out buds that extend in the head–tail direction until they meet, cross-linking the arteries. The new cross-links flourish, but all but the lowest of the original left–right links degenerate to leave the new cross-links making a pair of new arteries (vertebral arteries) that run in a head–tail direction. The survivors of the original plan, at the bottom of the figure, persist to become the subclavian arteries of the adult. There are many examples like this in the body, to the great dismay of most medical students faced with having to learn them.

running across the neck from near midline become united by what is effectively a new artery in the head–tail direction (Figure 44). Once this has happened, most of the original cross-wise arteries are lost, leaving just the new head–tail one (the 'vertebral artery'). This kind of re-modelling of the circulation happens a lot and it highlights the way that we have evolved our four-limbed, land animal's body by making a complicated series of modifications to the embryological foundations for a fish, rather than by starting all over again and building an embryo that is laid out like an adult human all along. It has probably been just too difficult (too improbable) to evolve mutations that radically alter the fish-like early embryo without breaking it completely, leaving small-scale, incremental tweaking of details the only feasible mechanism for evolutionary change.

The other 'new' method for making new blood vessels is suited admirably to adding elements to a system that is already up and running without breaking it, a serious constraint that was discussed in Chapter 1. It is called intussusceptive branching, and it creates new branches from a system in which blood is continuously flowing.[16] The process involves a zone of vessel wall at the top of the vessel, and another zone at the bottom, folding inwards until they meet. When they do, the cells re-arrange their adhesive contacts so that they join to one another, top to bottom, making a kind of pillar. The pillar can then extend along the vessel, effectively dividing what was one blood vessel into two parallel ones (Figure 45). The new centre wall can thicken by ingress by other cells of the

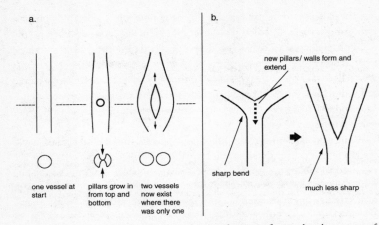

FIGURE 45 Intussusceptive branching. (a) Shows a plan view of a vessel at three stages of intussusceptive branching (top row), and a section through the vessel along the dotted line (bottom row). Ingrowths from the top and bottom of the vessel walls make a pillar, and this spreads longitudinally to divide the vessel in two. (b) Depicts how a similar process can alter the divergence angle of a branch point, even one initially formed by making a new side-branch, and hence smooth the flow of blood.

tissue, so that the two new vessels separate. Each can now do the same thing, making a total of four vessels where once there was one, and so on. In this way, one blood vessel can be transformed into a network without ever interrupting the passage of blood. Intussusceptive branching can also be used to move a branch point to make its angles of divergence less dramatic. This is important in smoothing out the flow of blood, a topic to which this chapter will soon return.

As well as making the main blood vessels of the body, and the heart to pump them around, the embryo has to make blood cells to travel in them. This involves a topological problem: blood cells should be found only inside blood vessels, and blood vessels form a closed circulation system with no entrance or exit to the rest of the body. How, then, can blood cells be brought into the vessels? The solution turns out to be simple: the first blood cells to form arise from cells that were once part of the wall of the aorta, and therefore have access to the inside of the vessel from the moment of its creation.[17, 18]

Making blood cells from the aorta wall may solve one problem, the topological one, but it creates another problem in that *some* of the aorta's cells need to decide to make blood, but they cannot *all* do so or the vessel itself would disappear. This requires an initially similar population of cells to divide into two groups, each of which will have a different fate and, once again, the decision-making process is guided by signals coming from nearby tissues. Tissues below the aorta, such as the gut, secrete signals that encourage vessel wall cells to take up a career making blood cells, while tissues above it, such as the neural tube, inhibit them. Some of the proteins involved have been identified, and the Hedgehog family, which triggered aorta formation in the first place, is involved yet again.[19]

While signals from surrounding tissues specify the general area of aortic wall in which cells might commit to making blood, other mechanisms are required to make sure that not all cells take this decision and destroy the wall. Cells in the region pre-disposed to make blood by the influence of other tissues begin to conduct short-range conversations between themselves. These conversations take place using molecules that are attached to the cells themselves, so that communication is intimate and neighbour-to-neighbour, rather than spreading to make a gradient. The details are still being worked out but some of the molecules have been identified and include the colourfully named proteins Notch, Jagged, and Mind-bomb.[20] Collectively, these short-range conversations ensure that not all the cells in that patch become blood cells.

The cells that leave the wall of the aorta are not yet blood cells; rather, they are stem cells for the blood line. The concept of stem cells was introduced in Chapter 3, where the cells in question were from such an early, uncommitted, stage of embryonic development that they could go on to produce any cell of the body. The cells that leave the aorta are a different type of stem cell: they have by now become committed to making blood cells of various sorts and that is all they will now make. The stem cells that come from the aorta move from the site of their origin, and take up home first in the liver, which is just starting to develop (Chapter 10). There they make all of the blood for the early foetus. Much later, when the bones have formed, the blood-forming stem cells move once again and settle in the bone marrow for the rest of their lives. Their story will be taken up again in Chapter 18.

The great vessels of the body are produced in a standard arrangement and do not vary much from individual to normal individual. In marked contrast to this, the small vessels and fine capillaries that bring blood deep into the tissues have a much more variable layout, developing according to the local needs of the tissues. This system has the great advantage that it can adapt to errors and changes in the precise positions and sizes of the tissues to be served. It depends on cells in the tissues issuing a biochemical cry for help if they feel themselves to be dangerously far from the nearest blood vessel.

The most serious consequence of a cell's being far away from circulating blood is a shortage of oxygen, a condition known as hypoxia. At least one of the systems that control blood vessel development makes direct use of this. Most cells contain a protein called Hypoxia Inducible Factor Alpha (Hif1α); in the presence of normal concentrations of oxygen, Hif1α is rapidly converted to a chemical state in which it is quickly destroyed by the cell's protein recycling machinery.[21, 22] Where there is little oxygen, though, Hif1α remains in a stable state and avoids being broken down and recycled. It therefore has time to enter the nucleus of the cell and to activate a small number of specific genes. The proteins specified by some of the genes have the effect of shutting down, temporarily, oxygen-hungry processes in the cell that are not strictly necessary for survival; in particular, they shut down cell multiplication to avoid adding yet more cells to an oxygen-poor area and making the problem worse. One of the other genes activated by Hif1α does something quite different: it specifies the production of VEGF,[23] a protein that encourages the proliferation and migration of blood vessel cells. This protein is the biochemical cry for help.

On detecting VEGF, cells in the walls of blood vessels, of both the artery and the vein type, begin to multiply and to reach out in the direction of strongest VEGF, which will be directly towards the tissue that finds itself short of oxygen. Branches of the small vessels grow into the tissues and, where they meet one another, they join to make a network of small tubes. The result is the production of a series of fine vessels—capillaries—that connect arteries and veins and bring fresh blood flowing through the previously hypoxic tissue.

When the blood brings oxygen to the cells that had been short of it, their Hif1α will rapidly be destroyed. With Hif1α fallen to insignificant levels, the genes whose expression was activated by it will no longer be expressed. The cells will therefore no longer be issuing any cry for help and they will be free to multiply. If they do multiply, then sooner or later there will be so many cells in the area that those that find themselves furthest from the new blood capillary will be short of oxygen and the whole process will repeat, the new capillary perhaps being one of those that responds by sending out yet more branches.

Systems like this ensure that, as cells multiply and tissues grow, the blood supply keeps up with them. The developing vessels do not need an accurate map of where and when each tissue will expand: all they have to do is detect cries for help from oxygen-starved tissues and go to them. Such a way of doing things is economical from the point of view of the information cells need to carry (much less than if they were following a map) and is almost infinitely adaptive since any tissue desperate for oxygen will be served.

The formation of new blood vessel branches is not only sensitive to signals coming from the tissues that the vessels serve, it is also sensitive to the flow of blood itself. Where a fluid flows down a tube, the slight drag between the moving fluid and the walls of the tube exerts a force on both. The fluid is slowed down, while the tube walls experience a force known as shear stress that runs along the walls. Smooth, gentle flow of fluid ('laminar flow', as fluid mechanics specialists call it) produces only gentle shear stress, and vessel wall cells that experience this react in a way that stabilizes their shape with no change. Turbulent flow, on the other hand, such as would happen if too much blood were surging through a small vessel, or a bend were too tight, exerts a strong shear stress on the vessel walls. Their cells detect this and begin to make new branches, either by sprouting, by intussusceptive branching, or both. The effect is very fast: clamping off one branch of an artery over the yolk of a chick egg, increasing flow elsewhere, causes intussusceptive remodelling of the arterial system to begin within fifteen minutes, and to continue until blood pressures and flow speeds are correct again, and the tissue is properly served.[24] Constant and automatic remodelling of the circulation system to avoid hotspots of turbulence results in a system of tubes in which flow is gentle, forces are low, and little of the heart's power is wasted by throwing energy away in turbulence. It is, once again, adaptive and automatic.

The adaptability of blood vessel development is very important to our ability to maintain and repair ourselves, as is explained in Chapter 18, but it comes with a price. Tumours, like ordinary tissues, consist of cells that are healthiest when oxygen is plentiful. When cells in the middle of a growing tumour begin to feel short of oxygen, they generally issue the same cry for help that would be given by normal cells of the body, and for the same reason. Having no way to know that the signal is coming from a potentially dangerous tumour rather than from normal cells in trouble, nearby blood vessels obligingly send branches into the tumour, giving it oxygen and nourishment and helping it to grow. One approach to developing anti-cancer medicines centres on the idea of blocking the ability of blood vessels to respond to a tumour's signals, so that the tumour will remain starved of oxygen and food.

The problem is complicated by the fact that cells have a variety of ways of crying for help, but reasonable progress is being made and, while the drugs being produced are probably not in themselves going to be a magic cure, intelligent use of them in conjunction with other treatments may significantly extend the lives of a significant proportion of cancer patients.

10

ORGANIZING ORGANS

It's organ, organ all the time... Dylan Thomas

A fundamental feature of large and complex animals such as humans is that their internal anatomy is not a continuous mass of cells but is instead divided into distinct organs, each specialized for a particular set of tasks. The lungs oxygenate blood, the thymus produces important defence systems, the intestines absorb nutrients from food, the pancreas makes enzymes to help the intestines digest food in the first place and makes hormones to regulate food use, the kidneys filter wastes from blood, and the uterus provides a home for the next generation. This separation of tasks into different organs is a way for the body to cope with running apparently conflicting activities at the same time. A child, for example, makes new muscle proteins to allow its muscles to grow and, at the same time, makes enzymes that digest the muscle proteins in the meat he has just eaten: clearly, these activities must be separated if a pointless cycle of synthesis and destruction is to be avoided.

The modular nature of a body constructed around individual organs is reflected in development, in that each organ builds itself using mainly local communication between its own cells. It refers to the rest of the body for only a few key decisions, such as when to begin building itself, when building should end, and how large to be. Internal details emerge mainly from internal processes. Indeed, many organs can be removed from their embryo and grown, at least for a few days, in an incubator, where they will continue to develop as they would in the embryo.[1]

The internal organs of the trunk can be divided into three main classes, according to how they develop. The first class contains just the heart, the very early development of which has already been described (Chapter 9) and which is unique in the way that it comes together. The second class is the largest, and contains organs such as the lungs, the liver, the pancreas, and the gall bladder. What these organs have in common is that they all develop as branches of the gut (Figure 46). The liver is the first of these organs to form, first in terms both of development and of evolutionary history. A branch forms from the wall of the gut, in much the same way that branches form from arterial walls (Chapter 9), and grows into the surrounding cells. The branch, which becomes the main drainage duct of the liver, then branches again, and its cells spread out to make a mass of liver cells. The liver cell mass then hollows out to make many fine tubes that converge on the drainage duct; the tubes will carry the liver's secretions to the gut where they will aid digestion. While all of this is going on, another branch emerges from the gut, and this gives rise to the gall bladder. Two more buds form close to the same site, and these each go on to branch again and again to make small trees which form the secretory system of the pancreas. Much higher up the body, towards the head, another bud pushes out from the gut. In fish, this gives rise to the swim bladder, a fairly simple bag that can hold air and control buoyancy. In mammals it is heavily modified: it branches, and then each of the branches begins to branch repeatedly to form a large tree-like structure. These trees become the airways of the lungs. In all of these cases, initial emergence of the bud from the gut is controlled by signals coming from the surrounding, mesoderm-derived cells. In all cases the construction of the tissues in the organ is a collaboration between endoderm-derived cells of the gut, which make the tubes in the new organ (airways, ducts, etc.), and mesoderm-derived cells that make the rest of the solid tissue. In some organs, neural crest-derived cells are also involved.

The third class of internal organs of the trunk consists of organs that come entirely from the mesoderm; these include the spleen, the gonads (which form in the trunk even in boys: they move down to the scrotum later), three pairs of kidneys (only one pair of which we keep in adult life), the uterus, and various tubes concerned with the plumbing of the urinary and reproductive systems. The development of most of these organs begins with the formation of two long tubes within the mesoderm, beyond the outer edges of the somites. These tubes, the Wolffian duct and the Müllerian duct, will form the internal plumbing of the male and female reproductive systems (Chapter 12). The Wolffian duct also produces

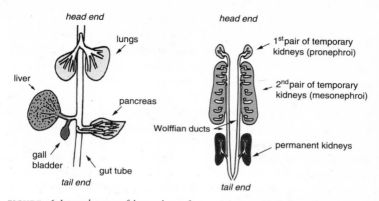

FIGURE 46 Internal organs of the trunk are of two main groups. (a) The lungs, liver, pancreas, and gall bladder all form from branches of the gut, itself derived from the endoderm. (b) The gonads, the temporary and permanent kidneys, the uterus, and the spleen all form from the same long block of mesoderm. The spleen, not shown here, also comes from mesoderm, but in a different place.

branches that are important in making the permanent kidneys (Figure 46). The main difference between the organs that come from the mesoderm and those that come from the gut is that mesoderm-derived organs such as the kidneys and gonads can make entirely new tubes as they go along, whereas gut-derived ones tend to rely on branching of the tubes they already have.

Each organ develops its own unique anatomy and uses its own specific sequence of developmental events to do so. The events themselves are drawn from a fairly small set of basic mechanisms such as making tubes, joining tubes, and clumping cells together.[2] Different organs often use exactly the same molecules to control these events. For this reason, this chapter will make no attempt to describe the formation of each individual organ but will use the development of just one, the permanent kidney, to represent the principles used by them all.

Before considering how kidneys develop, it may be useful to review their function and their final internal anatomy. The primary function of kidneys is to remove toxins and unwanted wastes from the body. There is an almost infinite variety of possible toxins that might have to be removed from blood, especially in an animal with a varied diet, so it is not feasible for the kidney to work by having specific export systems, each tailored for one specific waste. Instead, all small molecules, good and bad, are filtered out of the blood to a temporary holding area, then specific transport systems recover only the finite number of types of molecule that are still needed by the body. Filtration is performed by a few hundred thousand units called glomeruli, which are tight knots of small and rather leaky blood vessels that are covered with a fine filter and are located inside one end of a tube called a nephron. Each glomerulus has a nephron of its own. The fluid that leaks through the filter enters the nephron and proceeds along it, pushed by new fluid building up behind it. As the fluid travels, the cells that line the nephron grab and return specific substances, such as sugars, to the body. What is left behind drains into one of the branches of a tree-like system of collecting ducts, from which it passes to the bladder as urine. The main jobs of a developing kidney, therefore, are to make the urine collecting system, to make hundreds of thousands of nephrons that connect to it, and to lead blood vessels into the filtration ends of the nephrons where they can make glomeruli (Figure 47).

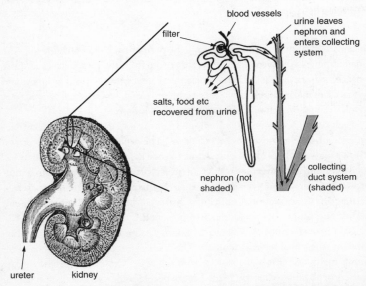

FIGURE 47 The adult human kidney, and a diagram of one of the hundreds of thousands of nephrons that it contains. One end filters water and small molecules, including wastes, from the blood. The folded tube recovers valuable small molecules, and the urine passes to the branched collecting system where, after removal of water, it goes on to drain to the bladder. The drawing of the mature kidney is from the first edition of Gray's Anatomy.

The kidneys are of the class of organs that develop from mesoderm. We form three pairs during our lives, although by birth only the third pair remains. The permanent kidneys arise near the tail end of the embryo's trunk. The first visible sign of them is the formation of a dense clump of a few hundred mesodermal cells. We still do not know exactly why this clump forms when and where it does, although it seems likely that the process is driven by a combination of the cells' mesoderm-specific proteins and by their expressing HOX proteins characteristic of this position along the head–tail axis of the body (Chapter 6). Researchers have identified a number of DNA-binding proteins that are required for the clumps of cells to form properly, but there is still a gap between identification of proteins and a true understanding of process.

Once it has formed, the clump of cells becomes known as the metanephrogenic mesenchyme (just called 'mesenchyme' in the rest of this chapter). Its first important action is to secrete a signalling protein, GDNF, that spreads out to nearby tissues. Just after this happens, the Wolffian duct, one of the tubes running down the trunk, sends out a branch tube directly into the mesenchyme (Figure 48).

The coincidence of these two events, production of a signalling molecule and production of a branch from a nearby tissue, suggests a causal connection. Several different types of experiment have shown that GDNF is indeed responsible for attracting the branch. In one type of experiment, GDNF activity was abolished, either by blocking it with drugs[3, 4] or by mutating its gene in mice:[5]

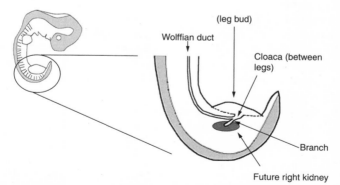

FIGURE 48 The first beginnings of the kidney: the shaded area of cells secretes the signalling molecule, GDNF, and the nearby Wolffian duct sends out a side-branch towards it.

without GDNF, the branch did not form reliably.* In another type of experiment, beads soaked in GDNF were placed near the Wolffian duct, and the duct responded by sending extra branches towards the beads, as well as the normal one towards the mesenchyme.[3, 4] These types of experiment, the first type showing that a signalling molecule is necessary for an event and the second showing that the pure molecule is a sufficient trigger, are typical of the way that researchers prove that a particular signal controls a particular event. They are typical of the work done to support statements about signals and responses that are made throughout this book.

This production of a branch by the Wolffian duct is directly analogous to the way that blood vessels send out a branch in response to VEGF (Chapter 9); the molecules are different (they have to be, to avoid confusion) but the action is the same. From this point, the organ, which consists only of this little tube surrounded by mesenchyme, can be removed from the embryo and placed alone in a Petri dish and it will continue to develop normally.[†] It is therefore already essentially autonomous and able to organize itself from this stage. Both the mesenchyme and the ureteric bud have to be present, though—neither can be cultured alone unless an experimenter goes to a great amount of trouble to 'fake' the signals that would be produced by the missing tissue.

As the mesenchyme signals to the tube that branched out from the Wolffian duct, so the tube signals back to the mesenchyme. Its signals cause the previously loose-packed mesenchyme cells to make a dense cap around the tip of the branch. Within this cap, the mesenchyme cells make new proteins and they begin to multiply. As they do, they continue to make GDNF and other signalling molecules, and these cause the branch to keep growing and dividing into more branches, each of which will divide again, and so on. As the tips divide, each inherits some of the cap mesenchyme. The result of this is the development of a small 'tree' of tubes, each bearing its own cap of GDNF-secreting mesenchyme cells (Figure 49). The tree of tubes will form the urine collecting system of the kidney.

A striking feature of this process is the inter-dependence of the branching tubes and the cells that cap them. In principle, one could imagine an alternative system in which each of these cell types would multiply of its own

---

* The branch (= ureteric bud) does form sometimes, even in GDNF$^{ko/ko}$ mice, suggesting some redundancy with other signalling systems.

† Except that it will not gain a proper blood system if just cultured in a dish. If it is cultured on the blood vessel-rich membranes in chick eggs, instead of a Petri dish, it can do even that.

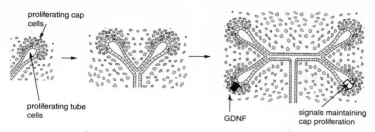

FIGURE 49 Mutual assistance between the cap cells and the branching tube. The cap cells secrete GDNF, which causes tube cells to proliferate and the tube to branch, and the tube cells secrete factors that maintain proliferation of the cap cells. Together, these signals keep the two populations in balance as the system grows.

accord. In such a case, however, there would be a real risk of one cell type multiplying faster than the other, so that the ratio of the two tissues would become more and more unbalanced and there would either be a lot of un-capped tubes or a large lump of cap cells with nothing to cap. Even if the relative multiplication rates in this imagined organ were exactly right, there would still be the risk that the tubes would extend in directions that take them away from the bulk of the cap cells, again resulting in a dysfunctional organ. In the real kidney, the dependence of the tube for cap-derived signals and the dependence of the cap for tube-derived signals keep the development of these two tissues properly in step. Any cells of one type that stray too far from an adequate population of cells of the other type will simply cease to multiply and will not create a problem. This inter-dependence between tissues is an important feature of the self-organization of the kidney and it is, unsurprisingly, seen in other organs as well. The identities of the signalling molecules are sometimes different, but the general principle remains the same.

In a fairly simple organ, such as the lung, a controlled pattern of branching like the one described above is sufficient to set up the main features of the organ: to a first approximation, lungs consist of a highly branched system of air tubes surrounded by loose-packed cells and blood vessels. The mature kidney, though, is more compli-cated because, as well as containing one highly branched drainage tube, it also has to contain the nephrons. When kidney development was first studied, it was assumed that the nephrons formed as side-branches from the branched drainage system. By the late Victorian era, however, it was realized that the nephrons actually form from the cap of mesoderm cells. This is an example of the habit of mesoderm-derived

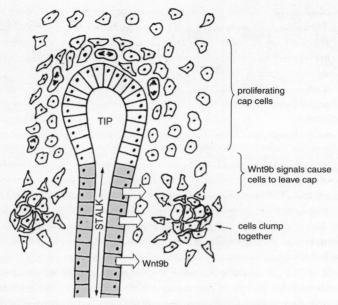

FIGURE 50 Cap cells that are left behind by the advancing tip find themselves near to the stalk region, which produced WNT9b. This, and perhaps other molecules, causes the cells to clump together in preparation for making a nephron.

organs of making new tubes from scratch. To understand how this is arranged, it is first necessary to examine the biology of the growing branches in more detail, because they are in control of what the mesenchyme does.

Within the branching tube, most cell proliferation is confined to a few tens of cells at the very tips of the branches.[6] The proliferation both maintains a population of tip cells and also gives rise to cells that will be left behind by the advancing tips to form the stalks of the tree. These stalk cells make a slightly different set of proteins, and among them is the WNT-type signalling molecule (WNT9b).[7] Those cells at the bottom rim of the cap, nearest the stalk, experience the highest levels of WNT, and this causes them to activate new patterns of protein expression. They leave the cap and form a small, tight clump of cells (Figure 50). This clump will go on to become a tube, the nephron.

By the time a group of cells is making the hollow sphere stage of a nephron, the branching drainage tube whose stalk triggered the whole process will have moved on. Its tip will have branched again and grown forwards, leaving new stalk behind, and new cohorts of cells at the stalk end of the ever-proliferating cap will have received a WNT signal from the new stalk. They will then form a nephron of their own. In this way, wave after wave of nephrons is formed as the drainage system advances.

The dependence of nephron formation on the WNT signal from a stalk means that nephrons only ever form close to the drainage tubes to which they will eventually connect. This is important for connections to be made reliably, and again it is a much safer system than one in which cells made their decisions independently of the tissues to which they must one day connect. Proximity to the branching ureteric bud runs the risk of bringing problems, however, because the construction of a nephron is a complicated and delicate business and would be ruined if a ureteric bud should send a branch through the area. This problem is avoided by cells that have begun to make nephrons switching off their production of the branch-promoting factor, GDNF, and instead making other factors that positively inhibit the formation of new branches in their vicinity. The risk of a new branch invading an area already making a nephron is therefore neatly avoided.

The switch of cells from making branch-promoting GDNF to making branch-inhibiting factors has larger-scale consequences. Cells that have not yet been approached by a branch of the drainage system will still make GDNF, but those in an area that has been invaded will not. This means that the formation of new branches will be directed, automatically, to areas that have not yet been invaded, and not to areas already served with existing branches. If, for some random reason, the drainage tree misses an area, that area will still make GDNF and continue to request a new branch until one has come. The branching tree therefore spaces itself out automatically, to serve the whole growing organ efficiently. The spacing may also be aided by mutual repulsion of branching tips. There is very strong evidence for such a system in developing milk ducts of the mammary glands,[8, 9] which also branch, and growing evidence in the kidney.

The kidney's blood system comes from the renal artery and drains to the renal vein—these form as branches directly from the aorta and the great vein (Chapter 9). Within the kidney, the vessels have to form a branching system to take blood into the hundreds and thousands of filtration units. Development of the kidney's blood system is controlled by yet more signalling molecules, one of which is the VEGF that featured in Chapter 9.[10] The end of the nephron that is specialized as a filter secretes VEGF and is attractive to blood vessel cells. Vessels therefore grow towards the filtration units and connect them properly to the blood supply. Presumably, there must be some system that switches off VEGF production once blood vessels arrive, so that new ones press on to find new filters, rather than all ending up trying to serve the nearest.

The development of the kidney, like all other internal organs, is therefore controlled to a very large extent by conversations between different cells, conversations conducted in the language of spreading signalling proteins. What each cell does, and importantly what signals it sends out, are controlled by the signals it receives. The system is therefore rife with inter-dependencies. This makes it very adaptable, both in terms of tolerance of error and in terms of being able to accommodate evolutionary change. If, for example, an evolutionary change in an animal makes it continue growing for longer, then extending the growing period of the kidney will automatically just add yet more branches to the collecting ducts and yet more nephrons and more blood vessel branches, with no need for any detailed changes to be made to the system. Major alterations in body size, and to a large extent shape, can therefore be accommodated by organs that organize themselves using internal conversations between their tissues. Were a defined 'blueprint' system to be used—even if such as thing were possible—then thousands of changes in detailed specifications would be needed to arrange a simple thing such as the doubling of an organ's size (indeed, for a blueprint that specified the location of each nephron, that would mean a doubling of the number of specifications in the blueprint). In fact, as far as anyone can tell, the number of genes used to make the tiny kidney of a mouse and the large kidney of a human is exactly the same.

The relative independence of developing organs, the independence that allows them to grow alone in simple culture systems, has another great advantage. Vertebrate evolution has featured the addition of a number of new organs.[11] The pancreas, for example, arrived with jawed fish, lungs with branched airways arrived with reptiles, and prostate and mammary glands arrived only with mammals. These organs often re-use control systems that had already evolved for an earlier organ. The branching of ducts in the prostate is regulated by FGF signalling proteins, which were already used for controlling branching of the tubes in the lung, and were used earlier still to control branching in the pancreas. The independence and relative isolation of the control systems within organs means that each can control itself without interfering with the others. Different organs can therefore re-use the same systems, again reducing the number of things that have to be invented anew in the evolution of new structures.

One surprising feature of kidney development, certainly in rats and probably in humans, is that it is highly sensitive to the nutritional state of the mother. If pregnant rats are given only half the amount of food they would normally eat, the kidneys of their pups have significantly fewer nephrons than normal. The effect of this is that body blood pressure is increased[12] (the chain of physiological events driving this is complicated, using feedback loops involving salt and hormones such as renin and angiotensin, but it can be thought of loosely as the body 'trying' to pass a normal amount of fluid through kidneys with too few nephrons by pushing the fluid harder). In humans, chronic high blood pressure is dangerous, elevating the risk of stroke, heart attack, and (ironically) kidney damage which will reduce the number of functional nephrons even more. It has been known for a long time that poor maternal nutrition can result in a child that will grow to have high blood pressure, chronic kidney damage, and other problems such as a predisposition to diabetes.[13] The environmental determination, during pregnancy, of later adult health, is called 'foetal programming'. It may well be that the great sensitivity of kidneys to maternal nutrition is a leading cause of the phenomenon.

From a well-nourished, Western European perspective, foetal programming seems to be purely maladaptive. It is arguable, though, that a reduced nephron endowment may have evolved to be an adaptation to periods of famine, allowing conservation of precious food molecules and salts in the blood. Two well-studied examples of population-wide maternal malnutrition, both dating from World War II, add some circumstantial support to this argument, although the data are about a different consequence of foetal programming, type II diabetes. In the Dutch Famine, an acute shortage of food for some months was followed by a time of relative plenty after the liberation of the Netherlands: children born to starved mothers went on to show, later in life, a high incidence of foetal-programming diseases such as type II diabetes. In the famine associated with the siege of Leningrad, there was no sudden relief and children were faced with meagre rations for a long period after birth. These offspring showed fewer ill-effects. The interpretation of these stories is contentious (the populations were not genetically identical, for example) but, in the minds of at least some researchers, they have suggested that the foetal programming may be positively adaptive for low-food conditions and the danger arises only when a baby adapted for famine actually meets a world of plenty.[13, 14] This in an important question, because it has clear clinical implications about whether one should feed low birthweight babies very high energy milk formulations to encourage rapid catch-up growth.

Every organ of the body has its own characteristic anatomy and its own particular mechanisms of development that are added to the general, shared ones. The take-home messages from this illustration of kidney development, particularly about the use of flexible communication in place of a 'hard-wired' programme can, however, be applied to them all.

11

TAKING UP ARMS
(AND LEGS)

A simple child,
That lightly draws its breath,
And feels its life in every limb...
William Wordsworth

Limbs are essential to the mammalian way of life. They are needed for move-ment, for defence and, in the case of humans and our closest relatives, for delicate manipulation of objects. In evolutionary time, vertebrates developed limbs long after they developed the basic anatomy of the trunk and head. Similarly, in developmental time, embryos develop limbs only after their basic body plan has already formed. In humans, the first signs of limbs appear around twenty-four days after conception, by which time the main structure of the trunk has been established and there is a working, if primitive, circulation.

Limb development begins with the emergence of two small projections from the sides of the embryo a little headward of the position of the heart. These will become the arms and, before long, two similar projections develop at the opposite end of the trunk to make the legs. The driving force for the emergence of these projections is an ongoing vigorous proliferation of cells just under-neath the ectoderm that covers the outside of the embryo. It is not that the rate of proliferation actually increases in these areas, but rather that it is maintained there even when growth of the rest of the embryo slows down: the net effect is that the developing limbs show higher proliferation than the rest of the flank.

These cells do not decide for themselves to maintain high rates of proliferation; rather, they are induced to do so by signals from the mesoderm of the trunk. Probably guided by their HOX code (Chapter 6), mesodermal cells at the levels of the trunk that will correspond to the arms and legs activate new gene expression,[1, 2] and one consequence of this is that they switch on the production of WNT family signalling proteins. The WNT proteins activate production of molecules of the FGF signalling family,[3] and these drive the production of a visible limb bud. The power of FGFs to cause the production of limbs has been demonstrated dramatically by researchers who implanted gels soaked in FGF protein or implanted viruses, genetically engineered to produce FGFs, into the flank of a chick embryo between the places that the wing and leg would normally form: in response, the overlying cells made extra projections, which went on to form extra limbs.[4, 5] Artificially driving WNT signalling in these sites has the same effect.[6] On the other hand, blocking the function of WNTs or of FGFs resulted in a failure of limbs to form even in their normal places.

Once these signals have initiated limb development, a thin stripe of ectoderm ('skin') on the limb-forming area thickens and the underlying mesoderm cells proliferate. This proliferation pushes out a paddle-shaped limb bud, the end of which is capped by the thickened ectoderm (Figure 51).

The cells in the bulk of the limb bud are, in themselves, capable of producing various types of tissues such as bones and tendons. Obviously, the cells must

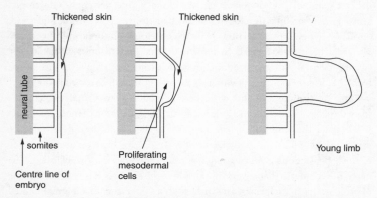

FIGURE 51 Outgrowth of paddle-like limb buds from the side of an embryo. The diagram shows the emergence of the right arm. The ectoderm thickens in a stripe where the limb will form, and proliferation of underlying cells makes a bump-shaped limb bud and then a paddle-shaped young limb.

not decide randomly what cell type to make when they mature or the limb would be a terrible mess; instead, they need to make a decision appropriate to where they are located. Those that will end up at the far end of the arm, for example, need to make the small bones of the fingers while those near the shoulder end of the limb need to make the humerus, the large bone of the upper arm. The other cells between them will make the elbow joint or the forearm bones, or the tendons and muscles appropriate to their place if they happen not to be along the lines where the bones form. The developing limb, therefore, requires a robust system that causes cells to make the appropriate tissue for their location. To the best of our current understanding—which is far from complete—they seem, like the cells of somites described in Chapter 7, to detect the concentrations of several different signalling molecules that each come from a different fixed place in the limb. Their fate may also be controlled by the passage of time.

The anatomy of a limb can be described according to positions along three axes: the shoulder-to-fingertip axis, the thumb-to-little-finger axis and the palm-to-back-of-hand axis. Although the systems that specify these axes are not completely independent in development, it will be simplest to consider them one by one and note the interactions when they arise.

At the time of writing this book, there is considerable debate about how position along the shoulder-to-fingertip axis is specified: there are several different ideas, each championed by its own set of adherents and none of them proved or disproved to the satisfaction of everyone concerned. At one level, this disagreement is a problem: this chapter cannot just give 'the answer' in any honest way. At another level, though, it is an opportunity to illustrate how biological science really proceeds.

First, some facts about which nobody disagrees: the limb elongates progressively, much cell proliferation taking place in a so-called 'progress zone' near its tip. As the limb elongates, the progress zone leaves cells behind it in the manner of a snail leaving a trail, and these cells go on to mature into the structures of the finished limb. Second, cells at the two ends of the developing limb experience different sets of (or at least, different ratios of) signalling molecules. The ectoderm cells at the very tip of the limb bud make FGF proteins.[7, 8] At the shoulder end, another molecule, retinoic acid (which appeared in Chapter 6 in the context of somite formation) enters the limb bud from the trunk of the body. Third, the precursors to the bones of the upper arm are visible before those of the lower arm, in turn visible before those of the hand.

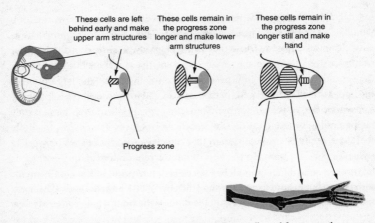

These cells are left behind early and make upper arm structures

These cells remain in the progress zone longer and make lower arm structures

These cells remain in the progress zone longer still and make hand

Progress zone

FIGURE 52 The timing model: as the limb bud lengthens, cells proliferating in the progress zone are left behind at different times. According to this model, cells are set to make structures closer and closer to the fingertip end of the arm the longer they have spent in the progress zone before being left behind. For clarity, the figure depicts just three states to correspond to three broad zones of the arm, but this model could operate at a much finer level of detailed specification than this.

One of the now-controversial models for the specification of limb position[9] is based directly on the fact that cells that spend different amounts of time in the progress zone end up in different parts of the arm. The model proposed simply that time in the progress zone was itself a kind of signal, and that the more time a cell had spent there, the more finger-wards character it would express when it matured (Figure 52). Cells that left the zone of proliferation early would mature appropriately for the upper arm, those that left later for the elbow region, those later still the lower arm, wrist, hand and, for those that never really left, the ends of the fingers. This elegant model makes the prediction that, if an experimenter could force cells to spend abnormally long in the zone, the resulting limb would be under-supplied with upper-arm tissue and over-supplied with hand tissue. An attempt to do exactly this was made by irradiating chick limb buds with X-rays, which killed many cells in the progress zone and caused all of the survivors to spend extra time there, making up the numbers again by multiplying, before some then left. The result was indeed a limb without a humerus but with the hand structures still formed, exactly as the time-based theory would predict.[10]

The problem with the experiment above is that the delayed exit of surviving cells from the progress zone was a very indirect effect of the X-rays. Their primary effect was killing other cells, raising an obvious concern that the lack of upper-limb structures may be explained by different cells having different sensitivities to X-rays. A recent re-visiting of this experiment[11] using modern techniques to examine gene expression showed that, in fact, the X-rays made no difference to cells starting out on the pathway to make upper arm. What instead happened was that cells maturing to make visible upper arm structures died under the X-ray stress. Thus an experiment originally thought to be a sound support of the timing model turned out not to be. The timing model is not destroyed by this, but rather left rather less supported by evidence than was once thought.

Another, quite different model for specification of position has nothing to do with timing but instead uses opposing gradients of FGF proteins from the finger-end and retinoic acid from the shoulder-end of the limb. The idea here is that spread of the signalling molecules means that cells at different points along the limb will experience different ratios of retinoic acid and FGF, and their future behaviour will be determined accordingly (Figure 53). Because the spread of the signalling molecules is limited, most versions of this model insist that cells must fix their identity early when the limb bud is small, even if their choice becomes visible only much later. The idea of the ratio-of-signals model is supported by experiments in which cells were removed from early limb buds and maintained outside the body in culture dishes, where pure signalling proteins could be added to them in such a way that all of the cells received the same concentration.[12] Addition of signals such as FGF proteins, which would normally come from the tip of the limb bud, to the cultures, caused cells to reduce expression of genes associated with the upper arm and to increase, first those associated with the lower arm, and then those associated with the hand. Addition of retinoic acid, normally coming down from the shoulder end of the limb, encouraged upper arm character even when FGF proteins were added at the same time. A broadly similar set of experiments in which limb bud cells have been grafted to other sites in the embryo, which happen to be rich in FGFs or retinoic acid, tell a similar story.[13] In particular, these experiments have made clear that making an upper arm requires retinoic acid or some equivalent signalling molecule spreading from the flank.[14]

If the ratio-of-signals model is correct, adding extra FGF to a limb bud ought to make it have a larger hand/finger zone at the expense of the rest of the arm. Unfortunately, this is not what is observed: instead, the zone of cells specified to be the hand is the same size as usual.[15] This suggests that the ratio-of-signals model cannot be entirely right, at least at the hand end of the limb.

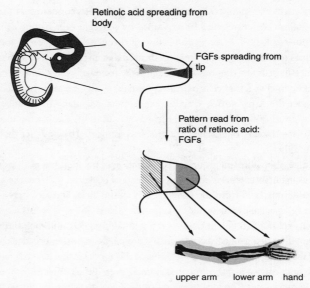

Retinoic acid spreading from body

FGFs spreading from tip

Pattern read from ratio of retinoic acid: FGFs

upper arm   lower arm   hand

FIGURE 53 The ratio-of-signals model: opposing gradients of FGF from the end of the limb bud, and retinoic acid from the body, activate different patterns of gene expression in cells so that those exposed to high FGF and low retinoic acid make structures of the hand, while those exposed to low FGF and high retinoic acid make structures of the upper arm. The precise levels of FGF and retinoic acid determine whether a cell makes structures characteristic of finger ends or the main body of the hand, or the shoulder or elbow end of the upper arm.

The current situation, then, is that we have at least two main models, each of which has some supporting evidence and each of which runs into at least one experimental result that suggests it is wrong. When this sort of problem arises in science, it can often be a good policy to try to identify any underlying assumptions that both models make. In this case, one hidden assumption made by both models is that a single mechanism patterns the entire shoulder-to-fingertip axis of the limb. This may be wrong. In our evolutionary history, limbs did not arise at once with all of their constituent parts. Lobe-finned fish, for example, have in their pectoral fins structures that are related to the upper and lower arm, but have nothing corresponding to the hand.[15] This suggests that the development of the hand was 'added on' later, a fact supported by land animal limbs showing the same basic sequence of gene expression as fish fins, and then adding on a whole new phase absent in fish.

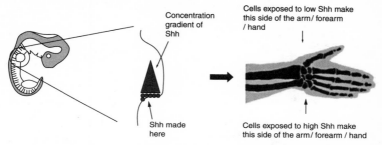

FIGURE 54 Cells use their local concentration of Sonic Hedgehog (abbreviated to SHH in the figure), produced at the little finger-side of the limb, to decide which type of finger or forearm bone (radius/ulna) to produce.

More primitive, jawless, fish have (or had, in the case of extinct species) even more primitive fins.[16] It is therefore possible that different parts of the limb, along what is now the shoulder-to-finger axis, are patterned by different mechanisms (Figure 54). Perhaps, for example, the ratio-of-signals model is a correct explanation of the specification of the upper arm—the flank derived retinoic acid-type signal may, for example, 'protect' the upper arm region from being made into anything else. For the lower arm and hand, though, beyond the range of flank-derived signals once the limb bud has elongated, perhaps the timing model is correct: this would explain why extra FGF failed to make the hand zone larger at the expense of upper arm.

Setting up the shoulder-to-finger axis is only part of the problem of setting up a pattern in the limb: there are also the other two axes to consider. Differences along the little finger–thumb axis of the limb are controlled mainly by a group of cells on the little finger side of the early limb bud. These cells make and secrete a signalling protein that spreads across the limb, its concentration weakening the further it spreads. Yet again, the signalling protein is Sonic Hedgehog. Cells in the hand that make and perceive a high concentration of Sonic Hedgehog early in their development go on to form structures associated with the little finger side of the limb, cells that receive less Sonic Hedgehog early in their development form middle fingers, and those that receive very little or none form the thumb (Figure 54). In the region of the forearm, the same applies, only this time cells experiencing receiving the highest Sonic Hedgehog make the ulna, and those that experience least make the radius and so on.

The idea that the type of finger built by cells depends on the amount of Sonic Hedgehog that they receive has been tested in a remarkable experiment

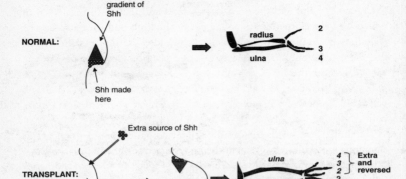

FIGURE 55 A normal chick fore-limb bud makes a limb with three fingers, equivalent to our 1st, 2nd, and 3rd fingers. If a second source of Shh is grafted into the opposite side of the limb bud to create two opposing gradients, the array of fingers is duplicated in mirror image and both forearm bones develop as an ulna.

done on a chick wing bud, the wing being the bird equivalent of a human arm. The experimenters grafted a second source of Sonic Hedgehog into an early chick limb bud, placing it on the 'thumb' side—that is, the wrong side. Both sides of the limb bud therefore had high concentrations of Sonic Hedgehog protein, and the lowest concentration was then in the middle. As the limb bud grew to form a wing, it made a second set of digits, but in the wrong order. From the normal 'little-finger' side of the wing, the sequence of digits was initially normal 1-2-3 in current nomenclature: normal chickens don't have a thumb or a fifth finger). Digit 2 was, though, formed near the middle of the limb where the Shh would have been at its lowest. Beyond it was a second digit 2, then another digit 3, and finally another digit 4. The whole sequence was therefore not the normal 1-2-3, but 1-2-3-3-2-1 (Figure 55). The formation of such an abnormal limb, exactly as predicted by theory, strongly suggests that the gradient-based mechanism outlined here is basically correct, although many details certainly remain to be discovered.*

---

* It is interesting to note that many human and animal mutations that result in the formation of extra digits on the thumb side of the hand have turned out to be in genes that control the production of Shh.

The third axis to be understood is the one that specifies the difference between the palm and the back of the hand and that is also reflected in which way the elbow hinges. Once again, the limb bud uses a source of signalling molecules that is localized to one side.[17] In this case, the signalling molecule is one of the WNT family, WNT7a, and it is made only on the side of the limb that gives rise to the back of the hand. WNT7a suppresses palm-type character: in mutant animals that have no functional WNT7a gene, both sides of the limb produce structures characteristic of the palm.[18]

The limb, then, seems to control its internal patterning by gradients of signalling molecules, the gradients being arranged at right angles to one another so that they cover all three dimensions of space. In terms of geometry, these axes are independent of each other but in terms of biochemical activity they remain tightly and confusingly connected.[19] If experimenters interfere with any one of them, the others also tend to fail, at least partially. The presence of the WNT7a signal that defines position along the back-of-hand to palm axis, is needed for the production of normal amounts of Sonic Hedgehog on the little finger side of the limb. The presence of at least some Sonic Hedgehog is needed to maintain FGF. Similarly, Sonic Hedgehog and FGF are needed for normal levels of WNT7a. We still do not know enough about limb development to understand exactly why there should be these inter-dependencies. It might be that they keep the signals coming from each of the sources in the correct proportion to each other and to the size of the limb.

As the limbs grow, they require the development of an efficient blood system to bring oxygen and nutrition to their multiplying cells. New vessels form in response to signals from the tissue, in the general manner already described in Chapter 9. The rapid elongation of the limbs, though, means that the blood system has to grow particularly quickly to keep up. If it were not to do so, cells of the limb would run short of nutrients and their multiplication would falter, leading to small, shortened, abnormal limbs. We know this from a profoundly tragic and completely accidental medical blunder that affected around ten thousand humans between 1958 and 1961.

The background to this whole incident was a laudable desire to control or eliminate the nausea ('morning sickness') that is a common and sometimes debilitating effect of early pregnancy. A drug that was already in use mainly as a tranquillizer and to control inflammation was found to be very effective in controlling morning sickness and was therefore prescribed to a large number of pregnant mothers. The drug was thalidomide. What was not known in 1958—and was in fact not known until this century—is that thalidomide breaks down in the body to produce a molecule that inhibits outgrowth of new and immature blood vessels.[20] This effect is strong enough that if the mother takes thalidomide during the time that fast-growing limbs are most in need of new blood vessels, the vessel growth can fall behind the limb, and limb growth fails. The ultimate result, in extremis, is the absence of limbs or small hands or feet apparently attached directly to the body. The reason why the hands and feet still form when the rest of the limb fails is not yet properly understood, although it does make sense in terms of the timing model if the lack of nutrients caused cells to spend far too long in the progress zone.

It took a few years before the sudden appearance of babies with deformed and shortened limbs in the population was connected with thalidomide but, by 1961, the connection was proven and thalidomide ceased to be prescribed for morning sickness. That said, it remains a useful drug against a variety of conditions including leprosy; its use in this application still causes an average of one or two birth defects each year worldwide. Thalidomide has also returned to the West, being very useful against some types of eye disease and cancer, precisely because it reduces the growth of blood vessels. Here, though, very great care is taken to ensure that it will not be given to anyone likely to become pregnant.

12

THE Y AND HOW

The reproduction of mankind is a great marvel and mystery.
Had God consulted me in the matter, I should have advised
Him to continue the generation of the species by fashioning
them out of clay.

Martin Luther

Everything up to this point has been about our shared experience, about the things that every one of us did back in the dreamtime of the embryo, before memory, before personality, before even the difference between male and female. For the first seven weeks or so of development it is impossible to look at an embryo and tell what sex it is, even by inspection of its internal organs; only an analysis of chromosomes will reveal that particular secret. The bodies of men and women arise from the same flesh and making the body parts unique to one sex or the other is not a matter of expanding anatomical differences that are there from the beginning, for there are none. It is instead a process by which some embryonic tissues follow one of two equally available paths of development. In mammals, the first parts of the system to make a decision between male and female are the gonads, and these then communicate their decision to the rest of the body.

The ovaries of women and the testes of men consist of a large number of different cell types which, for simplicity, can be divided into two groups. First, there is the germ line—the eggs and sperm themselves, and the cells that give rise to them. Cells of the germ line are the only cells capable of contributing genetic material to a new human and are, in this sense, the whole point of the reproductive system. Indeed some biologists see life mainly as a business of transmitting genes through generations and would argue that the germ cells are the point of the whole human: as Samuel Butler remarked, 'A chicken is an egg's way of making another egg'. The other group of cells in the gonad are known collectively as somatic cells, because they are of the existing body ('soma') and neither they nor their descendents can give rise to a new generation. The somatic cells include cells that make sex hormones, cells that protect and nurse germ cells as they develop into sperm or eggs, and various other cells that hold everything in place, organize a blood supply, connect the organ to the nervous system, and perform a host of other housekeeping tasks.

Somatic cells and germ line cells arise in different places in the embryo. The somatic cells of the gonad form each side of the midline in the upper part of the trunk. If you imagine your own body as the embryo, the gonads would be forming at about the level at which the lower parts of your lungs now lie. This location may seem utterly bizarre in the light of where gonads are found in an adult human, particularly a man, but it makes a kind of sense in the light of evolution. Humans, like all other mammals, reptiles, birds, and amphibians, evolved from fish. The gonads of even adult fish, particularly 'primitive' species, are located well forward in the body. At this location, they can interact with nearby tubes and other tissues to make a functioning reproductive system. These interactions are needed in humans too and, since the other tissues are also involved with making the aorta (Chapter 9) they cannot be moved without messing up the entire blood system. For this reason, the site of gonad formation cannot move either.

The germ line becomes identifiable when a group of around fifty cells is set aside in the epiblast at the posterior end of the primitive streak, just before gastrulation (Chapter 4). The cells probably go through gastrulation movements, but they do so very early and move out again with mesoderm that passes beyond the limits of the body itself.[1,2] When the processes of gastrulation that are described in Chapter 4 have made the trunk and head, the cells of the germ line are left outside to reside in the upper part of the yolk sac (Figure 56). Here they remain while the trunk of the embryo continues to organize its body and makes a neural tube, somites, circulation, etc. Once the basic body plan has been made, the germ cells then re-enter, propelled partly by cell movements involved in making the gut and partly by their own crawling. They use the outside surface of the developing gut tube as a kind of highway that takes them from the posterior end of the body up to the region of the developing gonads, into which they then crawl, being attracted by specific molecules. As they are doing all of

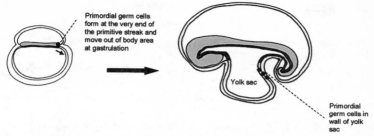

FIGURE 56 The cells of the early germ line ('primordial germ cells') arise at the posterior of the primitive streak and are carried outside the body proper, to wait in the yolk sac until the body has organized itself.

this, the germ cells also proliferate so that the original population of fifty becomes a population of about five thousand (and they will proliferate even more later).

Although the germ line will give rise to the sperm and eggs, it is not they but the somatic cells of the gonad that set up the sex of the body. The decision-making process uses a small set of interacting proteins. All have awkward names, but there is no way to present the mechanism of sex determination without naming some of the molecules: apologies in advance.

At this stage of development, a set of somatic cells begins to make a protein called WT1. WT1 performs a bewildering variety of functions in cells, but one of the most important is to bind specific sequences of DNA and, acting in concert with other DNA-binding proteins, to switch on particular genes. Most of the genes switched on by WT1 are located on chromosomes that are present in every human but one of them, *SRY*, is located on the Y chromosome.[3] Statistically, only half of all human embryos will possess a Y chromosome, for reasons that will be explained later. If the Y chromosome happens to be present in a given embryo, WT1 will cause the SRY protein to be made from it. If the Y chromosome is absent, then obviously SRY will be absent too. This makes all the difference.

Consider first an embryo in which SRY can be made. The SRY protein, like WT1, binds DNA, but it binds sequences of bases quite distinct from those that fit WT1. SRY switches on a new set of genes that would not be expressed in its absence. Amongst the genes activated by SRY is one coding for the protein SOX9, which activates the expression of yet more genes, so that the difference in the pattern of gene expression between a cell with a Y chromosome and one

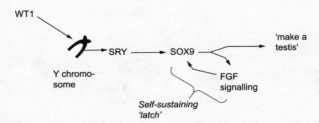

FIGURE 57 The 'maleness' cascade, that activates the Sox9-FGF signalling latch.

without builds like an avalanche.[4] SOX9 is not on the Y chromosome but on chromosome seventeen, present in every embryo.

One of the first actions of the SOX9 protein is to ensure its continued production. This is important, because the embryo has to make a clear and irrevocable decision about which sex to be if it is to avoid making a body that is neither fully male nor fully female. SOX9 expression activates a signalling pathway, based on the signalling by FGF proteins,* that in turn drives the production of more SOX9 even when SRY is no longer around.[5] The SOX9–FGF loop is therefore self-sustaining, and once it is triggered there is no way back: the embryo is committed to maleness (Figure 57).

The importance of the SOX9-FGF loop has been underlined by a series of experiments with genetically engineered mice. Mice in which *Sox9*† has been completely deleted in the gonad make female bodies even if Sry is present, because without a working Sox9, Sry has no way of influencing the body.[6] Deletion of the Fgf signalling system has the same effect. Conversely, mice in which Sox9 expression is forced on in the gonad whether or not *Sry* is present make male bodies, even if there is no Y chromosome.[7]

The genes expressed in response to SOX9 cause the SOX9-expressing cells to multiply strongly and to gain the shape and biochemical properties characteristic of the cells in the testis that support and nurture the production of sperm. Having made their decision, these cells organize all the other cells in the gonad to build a testis, the internal anatomy of which is a mass of tubes in whose thick walls sperm will one day be made by the descendents of the germ line cells.

---

* The proteins concerned are FGF9 and FGFR2. FGF9 is present anyway; SOX9 causes the cells to make the receptor FGFR2.

† The change from all capitals (SOX9) to mixed (Sox9) is just because of different conventions for human and mouse names.

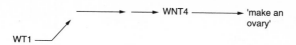

FIGURE 58 The 'female' pathway: WT1 expression leads, eventually, to expression of WNT4, which signals to cells of the gonad to make an ovary.

The paragraphs above describe what happens during male development. To follow female development, we need to go back to the time at which WT1 was expressed in a gonad that had not yet made any decision about which sex it would be. The *SRY* gene is not the only target of WT1, and a sequence of activation of other new genes, located on normal chromosomes possessed by all embryos, will naturally take place whether or not *SRY* is present. The details of the sequence have not yet been worked out fully, but it is clear that after some hours, cells make a member of the WNT family of signalling proteins that has already featured in Chapters 7, 10, and 11; in this case, the protein is WNT4. Provided that no strong inhibitors of WNT4 function are also present, the protein will cause cells of the gonad to develop into an ovary (Figure 58).

Obviously, this WNT4 pathway has to be stopped in males, and the FGF-SOX9 pathway, which is activated by the presence of SRY in male gonads, turns out to be a powerful inhibitor of WNT4 activity. The system therefore works by following a WNT4-driven, female pattern of development unless the male-specific genes are already expressed, in which case female development is suppressed.[8]

The male-promoting pathway used the SOX9-FGF 'latch' to make sure that cells that had started to follow the male route became really committed and would not waver due to random noise in the system. The female-promoting pathway that centres on WNT4 also uses a latch to ensure that any weak and random activation of male-promoting genes cannot mess things up. Once it is active, the WNT4 pathway strongly inhibits the male-specific pathway and ensures that no flickers of its activity could possibly end up setting the maleness latch in any cells. The choice of sex is therefore a struggle between two latches. If SRY is present and can switch the FGF-SOX9 latch in time, male-development is self-sustaining and the WNT4 pathway, necessary for female development, is held firmly 'off'. If SRY is not present or for any other reason fails to establish the FGF-SOX9 latch in time—about six hours in mice—then the WNT4 pathway becomes active, promotes female development, and holds any vestige of male development firmly 'off' (Figure 59).

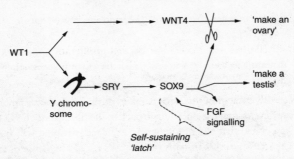

FIGURE 59 The molecular logic of sex determination. Expression of WT1 in the gonad begins a sequence of events that leads, eventually, to expression of WNT4, which signals to cells of the gonad to make an ovary. IF a Y chromosome is present, WT1 also causes the production of SRY from the Y chromosome: this causes the production of SOX9, which has multiple effects. It activates a signalling pathway based on FGF signalling that sustains SOX9 expression, it blocks the action of WNT4 (and therefore silences its 'make an ovary' signal), and it causes cells of the gonad to make a testis.

The importance of the WNT4 signalling system in promoting female development is illustrated by more experiments with genetically engineered mice. Mice in which the signalling by Wnt4 is strongly activated in the gonad by any of several genetic interventions develop female gonads and bodies, even if they have a Y chromosome.[9]

Cells whose WNT4 latch is in the female state develop into cells that support the development of eggs. They produce signals that make all of the other cells in the gonad make ovarian rather than testicular tissue (ovaries are not full of tubes, but full of loosely-packed tissue and clusters of cells that crowd around developing eggs). One of the signals made by these cells causes the germ cells to enter the special type of cell division, meiosis, that is critical to the formation of both eggs and sperm. In boys, meiosis does not begin until puberty but in girls all of the germ line cells begin meiosis almost as soon as the ovary begins developing: every egg that a woman can ever have is already in meiosis when she is born. Strangely perhaps, although meiosis begins before birth, it pauses for between about twelve and about fifty years, a few eggs re-starting their development each menstrual cycle. This pattern of development has a sad clinical consequence. Developing eggs paused in meiosis are very vulnerable to drugs sometimes used in chemotherapy, and girls and young women risk being made sterile by life-saving anti-cancer treatment. Fortunately, it is now possible to remove ovary tissue before chemotherapy begins, freeze it, and then to graft it back into the woman when she is of child-bearing age, should she decide that she wants to have children.[10]

Unlike the cell divisions used in normal cell proliferation, meiosis does not result in daughter cells that have all of the chromosomes that were present in the initial cell. A normal human cell, before division, has two copies of each chromosome, one inherited from that human's mother and one from the father. There are two versions of chromosome 1, two versions of chromosome 2, and so on. If the cell in question is female, there will be two versions of the X chromosome; if it is male, there will be only one X and one Y chromosome (because it is the presence of the Y chromosome and its SRY that makes a body become male, as explained above). While normal division effectively copies each chromosome so that the two daughter cells have the same constitution as the cell that made them, meiosis divides the chromosomes up so that one of the versions of chromosome 1 goes to one daughter cell and the other version of chromosome 1 goes to the other daughter cell. The daughter cells therefore end up carrying just one version of each chromosome, which is exactly what is needed in an egg or sperm, since the coming together of egg and sperm in fertilization will restore the normal total number of chromosomes again. Eggs, produced by meiosis in a female body, will all carry one X chromosome. In a male body, cells have one X chromosome and one Y chromosome. Half of the sperm made in that body therefore carry an X chromosome and half carry a Y chromosome. If an X-carrying sperm fertilizes an egg, the resulting embryo will have two X chromosomes and no Y chromosome, so will be female. Similarly, if a Y-bearing sperm fertilizes an egg, the resulting embryo will have a Y chromosome, with its SRY, so will become male. Our approximately 50/50 mix of men and women in the population[‡] therefore arises directly from the dynamics of chromosomes in meiosis.

Meiosis actually involves a little more than the partitioning of chromosomes. For each pair of your chromosomes, the version you inherited from your mother and the version you inherited from your father will be slightly different. This has nothing to do with your parents' sexes, but instead reflects their individuality. The DNA of chromosomes mutates over evolutionary time, so that

---

[‡] There are slightly fewer males than females, partly because males are vulnerable to lethal mutations on the X chromosome, having no second X chromosome to compensate, and partly because on average young males take more risks than young females (fighting, driving fast, etc.) so a greater proportion die young.

there are many subtly different versions of each chromosome across the human population. The variation in chromosomes is in part responsible for the variation in the colours, shapes, and abilities of different human bodies (the rest of the variation is due to environmental causes such as nutrition, disease, exercise, and culture). During the early stages of meiosis, the two versions of each chromosome come close to one another and 'recombine'—they effectively swap parts of themselves in a complex choreography of cutting and splicing DNA. This means that each version of a chromosome that is passed on to a sperm or egg is a hybrid of the versions in the normal body cells (Figure 60). Importantly, the details of which bits of chromosome are swapped in each cell entering meiosis are almost random, so that sperm and eggs made by the same individual differ from each other. That is why two children born of the same parents can be so different.

The recombination that happens during meiosis in a person's gonads is, effectively, the first time that that person's parents become truly united genetically. True, you carry chromosomes from your mother and your father, but these stand separate in each of your cells, giving contradictory instructions wherever they differ. In the language of gene function, your parents' chromosomes conduct a constant argument about how to bring you up. *'Blue eyes!' 'No, brown!' 'Tall!' 'No, stocky!' 'Calm.' 'No, easily stressed!'* and so on. Some traits such as eye colour are determined by relatively few genes, and others, such as susceptibility to stress, are controlled by a great many. Only when recombined in meiosis do genes contributed by your parents finally come together as one, ready to present a united front against the genes of your parents-in-law, when they embark on building the child you will call your own.

There is more to being a man or a woman than having the correct type of gonad. Most obviously, there are big differences in the rest of the reproductive system and, by the time development is over, there will be a host of subtle differences in other body parts too. By adulthood, these include the size of mammary glands (vestigial in men, developed in women), the shape of bones (the shape of the front of the pelvis is different in men and women, which is one of the ways that the sex of a skeleton can be deduced by archaeologists), the size of bones (on average, men are larger than women of a similar ethnic origin), the size of muscles (again, those of men being larger on average), the distribution of hair (men being more hairy), and the anatomy and function of parts of the brain. Cells in all of these body parts share the same chromosome constitution (XX or XY) as those in the gonads, but in mammals there is so far no evidence for any

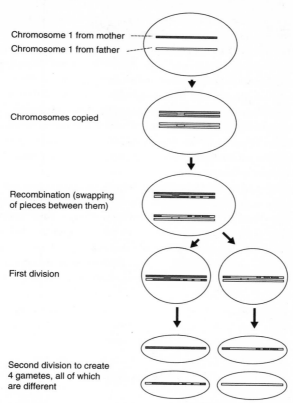

Chromosome 1 from mother

Chromosome 1 from father

Chromosomes copied

Recombination (swapping
of pieces between them)

First division

Second division to create
4 gametes, all of which
are different

FIGURE 60 Recombination between chromosomes in meiosis, in this case in a man's testis. For simplicity, this diagram illustrates only one pair of the twenty-three pairs of chromosomes in normal human cells. As for any division, the cell first makes a copy of all of its chromosomes. Then, uniquely for meiosis, it allows the version of the chromosome that was inherited from the man's mother to recombine with (swap pieces with) the man's father. Once this is over, two successive cell division events make four gametes, each of which has a different and new version of chromosome 1. The same will be true of chromosomes 2–21 (but not X and Y, the final pair, which do not recombine with each other). A very similar process takes place in a woman's ovary, except that one daughter from each division dies, leaving just one gamete from each starting cell, not four.

of these cells making the slightest attempt to make use of the fact.§ The Y chromosome, if present, seems to remain silent everywhere except for the early gonad. Instead of looking inward to their own chromosomes, cells of the rest of the body rely mainly on signals being sent from the gonad. These signals consist of small molecules easily transported over long distances: hormones.

If the gonads are removed from early rabbit embryos, so that no gonad-derived hormones of any type can be sent to the body, the embryos develop according to the female plan, whether they are XX or XY.[11] This experiment suggests that, left to itself, the non-gonad parts of an embryo will naturally build a female, and implies that hormones from the testis are needed to overcome this inbuilt female bias.

Once the gonad has committed itself, through the effects of the Sox9-FGF9-FGFR2 latch, to develop into a testis, it starts to secrete two important hormones; anti-Müllerian hormone and dihydrotestosterone. These mediate most of the effects of the testis on the rest of the embryo. Anti-Müllerian hormone gains its name from the effect the hormone has on a pair of tubes that run down the embryo, parallel to the Wolffian ducts that were mentioned in Chapter 10 (Figure 61). The Müllerian ducts are present in all early embryos and, if

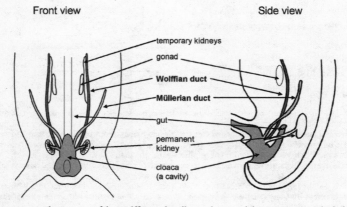

FIGURE 61 The position of the Wolffian and Müllerian ducts, and the structures to which they connect, in the hind part of the developing embryo.

§ A recent and surprising discovery has shown that even somatic cells of chickens pay attention to their chromosome constitution. This has fuelled speculation (so far, only speculation) that some mammalian cells might also do so.

they survive, they will develop into most of the internal plumbing of the female reproductive system. This includes the oviducts (fallopian tubes), the uterus, and the upper part of the vagina. Male bodies do not need any of these structures, and indeed they would get in the way of some aspects of male development. Anti-Müllerian hormone is harmless in itself but cells of the Müllerian duct make a set of proteins that will cause them to undergo elective cell death—a kind of cell suicide—if the anti-Müllerian hormone is detected. Elective cell death is a common and very important aspect of animal development and will be considered in more detail in Chapter 14.

Anti-Müllerian hormone, then, eliminates the precursors of the female internal reproductive system from a male body. The other half of the problem is constructing a male internal reproductive system. Here the Wolffian duct becomes important. A Wolffian duct runs on each side of the embryo, from the first temporary kidney, past the second temporary kidney and the developing permanent kidney, to open at the cloaca between the leg buds (Figure 61). This route takes it fairly close to the developing gonad. In the absence of testosterone, the Wolffian ducts and almost all of the temporary kidney structures die by elective cell death. In the presence of testosterone, the Wolffian ducts survive. They retain their connection with the temporary kidney, and a few of the tubes of this kidney grow out to invade the testis, losing any vestige of kidney function and instead becoming ducts for sperm transport. When the rest of the temporary kidney dies away, these remain and the Wolffian duct becomes the vas deferens** that, in adult life, carries sperm from the testis to the urethra, from which it can leave the male body.

If a gonad has become committed to forming an ovary, on the other hand, it makes neither anti-Müllerian hormone nor testosterone. In the absence of anti-Müllerian hormone, the Müllerian ducts survive and develop over time into the female internal reproductive system. Without testosterone, the Wolffian ducts die and disappear almost completely, eliminating the chance of developing male-type internal plumbing.

As well as having to develop male- or female-specific internal reproductive systems, growing embryos have to develop male- or female- specific external genitals. In contrast to the mechanism used for internal structures, which work by selecting which of two sets of progenitor tissues lives and which set dies, the mechanism used for external structures uses exactly the same founding tissues

** This is the tube cut in a vasectomy operation, a common method of surgical sterilization.

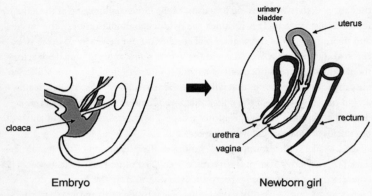

FIGURE 62 Division of the embryonic cloaca creates the separate openings of the urinary system, female reproductive system, and digestive system.

but sculpts them in different ways, rather as an origami enthusiast can make very different objects from the same piece of paper.

The external genitals arise from the tissues that are between the developing legs of the embryo, from about the fifth week of development. Exactly on the mid-line there is a cavity between the leg buds, called the cloaca: at this stage, it is a common opening for the gut, the urinary ducts, and the reproductive ducts (Figure 61). Its existence recalls the anatomy of our remote ancestors, such as fish and reptiles, in which digestive, urinary, and reproductive systems share a common outlet to the world. As time passes, the cloaca becomes divided by in-growing tissues into two or three separate openings. Its forward part becomes part of the bladder and its drainage tube (the urethra), its rear part becomes the rectum and, in females only, its middle part, to which Müllerian ducts connect, becomes the lower vagina (Figure 62).

During the fifth week of development, three swellings form next to the cloaca; there are two long ones running front to back, one on each side, and a third one where the long swellings meet just in front of it (Figure 62b). This last-mentioned swelling is called the genital tubercle. It grows out to become rod-like. In females it remains relatively small and becomes the clitoris, while in males it grows considerably to become the penis (Figure 63).

The swellings each side of the cloaca grow but remain in the same place in females; they become the labia, the paired lips that form the outer boundary of female genitals. In males, the absence of a vaginal opening allows the swellings

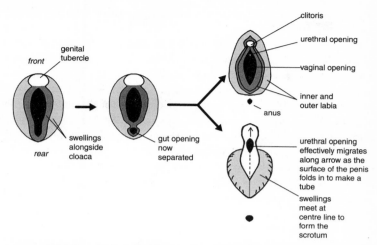

FIGURE 63 Making of male and female external genitals by different shaping of the same start-ing tissues. This diagram shows a developing human viewed upwards from between the legs. Initially, there is the opening of the cloaca and the three main swellings that surround it (ana-tomical pedants will divide some of these into sub-swellings, but grouping them into three is enough to illustrate the idea of what happens). After the cloaca divides to separate the anus from the urogenital system, the swellings develop, the genital tubercle extending to become the clit-oris or penis, and the other swellings either remaining where they are to become the labia, or closing over the mid-line to become the scrotum.

to meet along the mid-line and, when they meet, they join to make a continuous surface, the scrotum. The testes eventually descend into the scrotum by the action of a very long ligament that extends all the way from the gonads to the developing genitals. The ligament pulls the testes to their final positions at first just by not growing (so the rest of the body, which continues to grow, 'out-runs' the testes and leaves them in lower and lower positions), and then by actively shortening to pull them right the way down to the scrotum. In female embryos, ovaries are also each attached to a ligament but this shortens much less, so pulls the ovaries down into the pelvis, but no further.

The process of making a body become either male or female involves many steps and rather complicated developmental machinery. It is perhaps therefore not surprising that it can fail. Unlike failures of other complex developmental events, such as making the heart, failure of sex determination is not life-threatening, so plenty of humans are born who are in the 'wrong body' for their chromosomal sex, or who have bodies intermediate in character between male and female, or who have some body parts completely male and others completely female.

Mutations that stop either SRY or SOX9 working properly cause an individual with an XY chromosome constitution to make an essentially female body. Humans with inactivating mutations in SOX9 also show a type of dwarfism called campomelic dysplasia because, unlike SRY, SOX9 performs important duties in the development of parts of the body that have nothing to do with sex.[12] Mutations that result in extra copies of SOX9 being present on the chromosome have the opposite effect on sex: in 1999, medical researchers described the case of a small boy who had a basically male body, but he had two X chromosomes and no Y.[13] He did, though, have a duplication of a length of chromosome seventeen that included SOX9, which was presumably therefore made in high enough amounts, even with no SRY, that it could activate the SOX9-FGF latch.

The fact that gonads assess their chromosome constitution and communicate their findings to the rest of the body via hormones creates the opportunity for an interesting set of errors that create gonads characteristic of one sex and a body characteristic of the other. Testosterone, the messenger of maleness, can be detected only by means of its receptor. If this receptor is inactivated by mutation, the body will not detect the hormone and will therefore follow a female path of development even though its gonads are testes. Many individuals with this condition are indistinguishable in appearance, even when naked, from normal women (although they will not have menstrual cycles as they have no ovaries, and they will not have Müllerian-duct-derived structures because anti-Müllerian duct signalling will still function).

One of the most extraordinary abnormalities of the male/female choice seen in humans is caused by a mutation that is rare in general, but relatively common in some small island populations. The mutation affects the processing of testosterone in the tissues. Although it is regarded in popular culture as a synonym for masculinity, testosterone itself is actually quite a weak masculinizing hormone. It is converted by an enzyme in the tissues into a closely related but much stronger hormone, dihydrotestosterone. When the gene specifying that enzyme is mutated, testosterone is left unconverted: it is too weak to switch the body into masculine development and affected individuals can be born with apparently female bodies, as judged by outside appearance, and live their early lives as girls.[14] Years later, the onset of puberty brings, as it does in normal males, a burst of testosterone production. There is now so much testosterone in the system that, weak as it is, it is enough to drive masculine development. This can be enough to cause the phallus, initially clitoris-like, to grow into a penis, for the testes to descent into the scrotum, and for male patterns of hair growth to develop. The person who had lived the childhood of a girl now lives the rest of his life as a man (like Virginia Woolf's *Orlando*, but in reverse).

Less extreme abnormalities of signalling and response create milder ambiguities in the sex of the body. One individual with such an ambiguity made a significant, if unintentional, contribution to our understanding of sexual development. Inés de Torremolinos, who lived in the sixteenth century, was a widow who had three children, so she must have been functionally female but, perhaps because of excess testosterone production probably caused by an endocrine disease, she developed a condition now called cliteromegaly. In this, the clitoris is unusually enlarged and has a form intermediate between that of a normal clitoris and penis. Her physician, the doctor/scientist Mateo Columbo, noticed the unusual organ and understood, by its intermediate form, the connection between the development of male and female genitalia. This is probably the first time that it was understood that male and female external genitalia are two versions of the same thing and not completely different structures with different origins. He conducted 'therapy' on this patient and then on normal women, 'therapy' of the kind that today would see his being rapidly struck off from the General Medical Council List of Registered Practitioners, and realized through his 'work' that the clitoris was sexually sensitive. His publication of the work in 1558 is therefore also widely credited as the 'discovery' of the sexual function of the clitoris. This seems a bizarre claim, however, given that Roman writings, verging from classical poetry to obscene graffiti in the ruins of Pompeii, make it perfectly clear that the sexual sensitivity of the organ was known to men thousands of years ago: one assumes that the female half of the population must have been aware of it forever.

Some failures of complete sexual specification can be caused by the environment rather than by genetics. The fertility of Western males has been falling, on average, over recent generations, and there is mounting evidence that this is, at least partly, because environmental pollutants interfere with hormonal signalling.[15, 16, 17] Phthalates, used to increase the flexibility of plastics, are a particular concern, and significantly inhibit male development in laboratory animals. There is also a correlation between exposure of mothers to phthalates and development of less completely male genitals in their sons. Phthalates have recently been banned in toys in the European Union, though there are plenty of other compounds that still cause concern. The whole issue is controversial and is inevitably clouded by politics. Whether or not it turns out that current levels of pollution have a significant effect on human reproductive function, the animal data should at least act as a warning that, robust as development is, we cannot take for granted that it can survive everything that we may carelessly throw at it.

13

WIRED

Only Connect!... Live in fragments no longer.
E. M. Forster

O f all of the organs that the human foetus develops, the central nervous system is surely the most remarkable. When finally complete, it consists of tens of billions of individual cells, each of which can be connected so densely that a single cell can be connected to well over a thousand others. There are areas of the human brain in which there are around one hundred million connections in every cubic millimetre. For comparison, the microprocessor at the heart of the computer on which this book is being written contains far fewer connections than the brain. If we compare the number of nerve cells (neurons) in the brain with the number of transistors in that chip—a very conservative comparison because each neuron is, in terms of its function, more like a full microprocessor than a simple transistor—we find that the brain has about three million neurons for every transistor. In cooperating to develop such an organ, the simple cells of the embryo have to achieve feats of assembly that are massively more complicated than those we have so far achieved with our most advanced engineering. Though we have very much more to learn about how they do this, recent decades have produced at least the first glimmerings of an understanding.

The central nervous system develops from the neural tube of the early embryo, which is, in turn, formed by the in-folding of the neural groove along the back of the embryo (Chapter 5). The part of this tube that will be in the future head enlarges to become the brain and the rest becomes the spinal cord. Once the basic tube of the central nervous system has formed, the next phase of its development is dominated by cell proliferation. This runs faster than would be needed for the length of the neural tube to keep up with the general growth of the embryo, and this leaves 'extra' cells that are used to thicken the walls of the tube. The process of wall thickening involves a somewhat complicated choreography of cell movements, and movements of nuclei within cells, but the result is simple enough: production of a series of layers. Broadly speaking, the left and right sides of the neural tube are thickest, while the very top and the floor are thin.

Previous chapters have already considered patterning along two dimensions of the neural tube. Patterning along the tube was achieved by the chromosome-opening processes that activated different combinations of HOX genes at different levels of the head–tail axis (Chapter 6). Patterning across the whole tube was achieved by gradients of signalling molecules that spread up from the floor plate and down from the roof to instruct cells to develop into different types according to their position on that axis (Chapter 7). The formation of layers in the side wall of the neural tube adds a third, radial, direction of patterning (Figure 64). Cells finding themselves in a layer next to the cavity at the centre of the neural tube can become different from the ones in the middle of the wall, and these in turn become different from the ones at the outside. In some parts of the central nervous system, such as the cerebral cortex, the structure is very clearly layered in a way that is important to brain function, whereas in other places the layering is less obvious.

While the axes used to pattern the neural tube are conceptually simple, the situation becomes complicated by the fact that not all cells remain where they

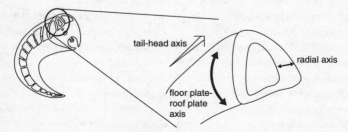

FIGURE 64 The three axes along which neural cells can detect their position and specialize accordingly. The left diagram shows the trunk of the embryo (head not shown), including its neural tube and the somites that flank it. The right diagram shows an expanded view of the neural tube with the three axes marked.

are created. In the brain, in particular, substantial numbers of neurons migrate within the plane of the neural tube and move from one place to another. Many of the neurons in the cerebral cortex, the part of the brain broadly concerned with what we normally mean by 'thinking', do not arise within that tissue but migrate there from other parts of the developing brain. Similarly, many cells in the olfactory bulb—the part of the brain connected with the sense of smell—come from other parts.[1] The cells navigate using the types of mechanism described in Chapter 8, and many of the guidance molecules have now been identified (although much remains to be discovered).

Neurons connect to each other mainly by sending out fine processes: dendrites, which receive the neurons' inputs and can perform some local calculations, and axons, which carry the output of a neuron on to other neurons or muscles. Axons can extend to be tens of thousands of times longer than the diameter of the cell body from which they arise. The axons that connect the base of your spinal cord to your feet, for example, are nearly a metre long, yet the cell body from which they project is only about ten thousandths of a millimetre across. Being very long, fine projections that exist to connect things, axons can be thought of as the 'wires' of the nervous system. As in electrical engineering, they have to travel large distances, so the body's 'wires' generally do so as a cable (nerve) containing hundreds of individual processes. The analogy is strengthened by the fact that the signals carried by axons are electrical in nature, although in axons the currents flow in a more complicated way than they flow in a wire, and the analogy should not be pushed too far.

Where axons meet another cell body, they form a special junction structure, the synapse, through which their signal is passed on. In some places, synapses involve a direct electrical connection to that cell. More commonly, they involve the axon releasing a small molecule called a neurotransmitter, which spreads across the small gap between the axon and the other cell and stimulates receptors on that cell. These receptors then activate electrical and/or biochemical activity in the receiving cell, and communication is achieved. Different types of neurons use different neurotransmitters. It is these neurotransmitters that are mimicked or inhibited by drugs—legal and illegal—that affect brain function by enhancing or depressing the activity of some brain systems while leaving others untouched.

The main problem faced by the developing nervous system is to arrange all of this 'wiring' so that neurons are connected properly to one another and, where applicable, to organs of sense (eyes, ears, nose, touch, temperature, etc.) or to

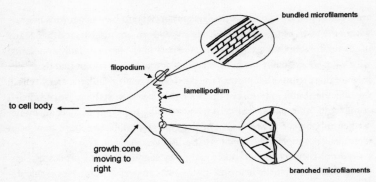

FIGURE 65 Microfilaments of the growth cone can be arranged as two protrusive structures, the lamellipodium (leading edge), supported by a branched network of individual microfilaments, and filopodia which are pushed out by the growth of bundled microfilaments.

organs of action (muscles, blood vessels, glands, etc.). Much of the work needed to achieve this is done by special structures found at the ends of growing axons—the growth cones.

Growth cones (Figure 65) consist mainly of the proteins that run the cell migration mechanisms described in Chapter 8.[2] They have a leading edge at which microfilaments are organized into a protrusive network, their fibres growing and pushing the membrane forward at the front. This pushing forward can sometimes include the formation of long, thin, spike-like filopodia that reach far forwards of the growth cone before retracting again. Further into the core of the growth cone, motor proteins like myosin interact with microfilaments and make them into contractile bundles that pull back on the leading edge. If there were nothing to resist this pull, the leading edge of the growth cone would simply retract and no motion would take place. To prevent this, the growth cone is equipped with complexes of proteins that adhere to specific components on the surface on which it crawls.[3] These protein complexes effectively anchor the microfilament systems and provide something against which the leading edge can push. They also stop the myosin of the central zone of the growth cone pulling the leading edge backwards: instead, the pulling has the result that the bulk of the growth cone is hauled forwards towards the leading edge. By this means, the growth cone, and therefore the axon behind it, advances.

It follows from the mechanism outlined above that there is an intimate connection between the ability of growth cones to adhere to a surface and their ability to advance. This is particularly important where the growth cone spans a choice of surfaces, either in a laboratory culture dish or in a living embryo. If the surface underlying one side of the growth cone offers more adhesion than the surface underlying another part, the centre of the growth cone will advance more towards the stickier side. The stickier side will also be more able to push the new leading edge forward. The growth cone, therefore, steers in that direction.

Different types of neuron make different adhesion complex proteins, and even one single type of neuron can produce different sets of adhesion complex proteins at different stages of its life. Each of these complexes sticks to a different partner molecule on an underlying surface, provided that the partner molecule is present. In this way, it is possible for different neurons to be faced with the same choice of surfaces but to make opposite decisions about where to grow.

Differences in adhesion are not the only navigational cues that can be read by growth cones. Other guidance cues work by signalling to the molecular machinery that assembles the growth cone's leading edge.[4] If different areas of the growth cone experience different local concentrations of external signals, the balance of protrusion and retreat will be different between these areas, and the growth cone will be steered in the direction of signals that activate protrusion and away from ones that activate retreat.[5] Sometimes, the differences in signal concentration between one place in the embryo and another are so strong that they produce an all-or-none response in the growth cone, excluding growth cones of a particular type completely from one area. In other cases, the differences in concentration are less extreme and the response of different growth cones is a relative one that can be used to match them to their destinations in a subtle way. In other cases still, there is a gradient of concentration that draws growth cones towards a distant target.

An example of the all-or-none type of response can be seen in a system that determines whether axons cross the midline of the spinal cord. Strong control of this is very important for a person to be able to move asymmetrically, for example to be able to hold a saucer still with the left hand while raising a teacup with the right. Raising the teacup involves voluntary contraction of the biceps muscle (and others). To the best of anyone's knowledge, there is nothing intrinsically different between the molecules expressed by the biceps of the left arm and the right arm. Growth cones of motor neurons in the spinal cord that are destined to control biceps would not in themselves be able to tell the difference. If growth cones could wander freely across the midline of the spinal cord before exiting it to seek the muscles, many of the neurons that should control the right arm would control the left one too, and vice versa, so the arms could only move together. It is therefore very important that this cannot happen, and motor neurons on the right side of the spinal cord can reach only the right arm. The same argument applies to sensory systems: we can hear whether someone calling us is standing to our left or right because the brain connects to the sensory systems of our left and right ears with a precision that allows them to be distinguished. It is true the body can detect and correct for occasional errors, by the processes of refinement that will be described in Chapter 15, but these will only work if most of the wiring goes to roughly the right place in the first instance.

Whether or not a growth cone, say of the axons of an interneuron that passes signals from one spinal cord neuron to another, is allowed to cross the mid-line depends mainly on its reaction to contact with cells of the floor plate, the strip that runs along the ventral surface of the neural tube (Chapter 5). Floor plate cells display on their surfaces a protein called SLIT. SLIT is detected by a receptor called ROBO, which is present in the growth cones of some neurons. When ROBO binds SLIT, it triggers signalling pathways inside the growth cone that block protrusion of the leading edge and also cause it to be pulled back actively.[6] A part of the leading edge of a ROBO-carrying growth cone that blunders into contact with a cell expressing SLIT therefore collapses, blocking the advance of the axon in that direction. The axon will veer off in the direction of some part of the leading edge that has not contacted the SLIT-carrying cells of the mid-line, and it will therefore never be able to cross. If a growth cone does not possess ROBO, on the other hand, that growth cone is completely blind to the presence of SLIT and can cross the midline with impunity (Figure 66).

Growth cones can change the set of proteins that they express at different stages of their lives. The axons that are supposed to cross the midline are an excellent example of this. On their way down to the midline, they possess receptor proteins that detect attractive signals that are released by midline cells and are therefore attracted there, and they express almost no ROBO and make proteins that dull the effect of ROBO signalling anyway, so they are able to cross the midline. As they are crossing, though, they are exposed to the high levels of Sonic Hedgehog that are made at the floor plate (Chapter 7), and this causes them to become sensitive to ROBO after a small delay that is just long enough to allow them to complete their crossing.[7] Their new sensitivity to ROBO means that the growth cones now find the midline repulsive: they therefore do not attempt to re-cross it but instead move away to travel on to their final targets.[8] (The name

a.   b.

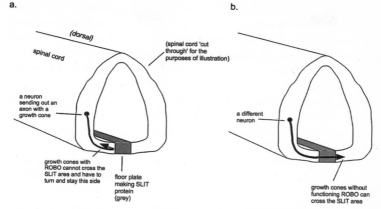

FIGURE 66 The floor plate of the spinal cord contains the signalling molecule SLIT. This is repulsive enough to growth cones that make functioning ROBO (a) to stop them crossing the area, while those with no functioning ROBO (b) are free to cross.

'ROBO' comes from the fruitfly mutation *'roundabout'*, which inactivates the fly version of ROBO and results in growth cones crossing and re-crossing the midline in small circles, as if they were vehicles driving around a roundabout.)

The axons of cells that are only supposed to connect on their own side of the central nervous system produce axons that must not cross the floor plate. The growth cones of these axons express ROBO and are repelled by contact with the floor plate. They therefore cannot cross the unwelcoming territory. Some of these neurons are sensitive both to the attractive signals that emanate from the floor plate and to the repulsive SLIT at the floorplate. Faced with the conflicting influences of attraction and repulsion, they stay as close as they can to the attractive signal, without being so close that repulsion becomes too strong.[9] Like moths flying as close as they can to a flame without being burned, they maintain a constant distance from the floor plate as they grow parallel to it, along the length of the spinal cord. These connect different levels of the body to one another along the head–tail axis, for example connecting the brain to the motor neurons at the level of the arms that control the biceps of someone lifting a tea cup.

An example of repulsion operating in a relative, rather than an all-or-none, way is seen in the way that the eye is wired to the brain. The mature eye works by focussing an image of a scene onto the retina, which is a curved 'screen' that lines the back of the eyeball. The retina is covered in light-sensitive cells, which alter the voltage on their membranes according to the brightness of the light that impinges on them. The light-sensitive cells are connected to neurons in their immediate vicinity that perform some local processing of signals before passing them on to the brain. The processing cells each send out an axon that travels directly into the brain. Many go to an area near the back of the head called the superior colliculus* in mammals. The axons of the retinal ganglion cells all run parallel to each other as a thick cable—the optic nerve—but when they reach the superior colliculus they disperse and connect to it in a quite remarkable manner: the place to which each one connects to the colliculus depends, in a precise way, on each ganglion cell's place on the retina. In effect, the exact layout of the ganglion cells on the retina is replicated on the colliculus so that it has a fully laid-out image, in electrical activity, of the optical image that is present at the back of the eye.

Several mechanisms cooperate to create this mapping of retina to colliculus, and turn an initially rough mapping into a more precise one. One powerful mechanism is provided by the interaction of repulsive molecules carried by cells of the colliculus with receptors on the growth cone that, if they detect the repulsive molecule, encourage local collapse of the growth cone's leading edge.[10] The production of the repulsive molecule on the colliculus is not uniform. Instead, it is very strong in the part of the colliculus that should end up being connected to axons from the side of the eye nearest the nose, the 'nasal side'. It becomes progressively weaker across the colliculus until it is very low in the part that should end up being connected to axons from the side of the eye that is nearest to the ear.[†] The production of the receptor by the growth cones of retinal ganglion cells is also not uniform: the growth cones coming from the nasal side of the eye make very little, but the amount expressed rises steadily until the growth cones coming from the side of the eye nearest the ear express very large amounts. The growth cones from the ear side of the eye would be repelled very strongly from colliculus cells that have significant amounts of the repulsive protein; they therefore steer strongly to the edge of the colliculus that has least. Growth cones nearer the middle of the eye still express some receptor, so still try to veer away from colliculus cells expressing the repulsive molecule.

---

* Technical note: the superior colliculus is the name given to this area in mammals: the equivalent in birds is called the optic tectum. The material in this section of his chapter is drawn from work in both types of animal, but only the mammalian word is used for simplicity.

† In technical language, it is usual to refer to the nasal and temporal sides of the retina, the temporal being a bone in the skull.

Space on the colliculus is limited, and these cells from the middle of the retina cannot compete with the determination of the growth cones from the 'ear' side of the eye, that have the highest amount of receptor, to connect to the part with the very lowest amounts of repulsive molecule. Instead, they have to settle for the areas that are modestly repellent. Growth cones from the nasal side of the eye, on the other hand, carry so little receptor that they are able to tolerate even the part of the colliculus that has very high levels of repulsive molecule. By their competing to get away from the repulsive molecule, and competing harder the more sensitive they are, the growth cones therefore arrange themselves on the colliculus in the same spatial order that they left the retina (Figure 67).

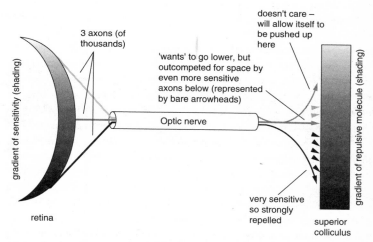

FIGURE 67 A representation, using just three axons from thousands actually present, of the mechanism by which the 'map' of the retina is re-created on the superior colliculus. Axons originating on the side of the eye nearest the ear, shown at the bottom of the diagram, carry a great deal of receptor and are therefore strongly repelled by the repulsive molecule on the colliculus. As this is present in a graded manner, they are strongly unwilling to make contact anywhere but at the end that has least (the bottom, on the diagram). Axons originating in the middle are also repelled, but less strongly. They therefore settle for the middle of the colliculus, which does not produce too much of the repulsive molecule (they would go lower, but the axons from the nasal side outcompete them). Axons originating from the nose side of the retina carry little receptor and are therefore not bothered by a repulsive molecule they can barely detect: they therefore connect to the part of the colliculus richest in repulsive molecule, space for which no other axons compete. It must be stressed that this figure is very diagrammatic and makes no attempt to represent the complex pathway of a real optic nerve, or the shape of the colliculus.

The mechanism described above orders growth cones according to their place along the horizontal, nose–ear axis of the eye. A very similar system that uses other pairs of repulsive molecules and receptors organizes the brow–cheek axis, so that the 'wiring' of the eye to the colliculus is a faithful two-dimensional map. For both axes, it is likely that there are still-undiscovered signalling systems, perhaps attractive ones as well as repulsive ones, that act with the repulsion-based mechanisms described above to refine the map. We are still in the very early stages of understanding all of this.

The account presented above, about how growth cones from the optic nerve connect to the right place in the brain, is mainly a story of very local interactions. It took for granted that the growth cones would already have made their long journey from the retina to the colliculus. That first part of the journey is itself complicated, and it uses a variety of repulsive and attractive cues.[11]

The first navigational challenge in the path from eye to brain is for a growth cone from the retina, which could be anywhere in the retina, to find the point at the back of the eye from which the optic nerve should emerge. This is achieved by the growth cones having receptors for repulsive molecules that spread from the edge of the retina, and their having receptors for attractive molecules that are located towards the centre of the retina from which the optical nerve will emerge (Figure 68). One of these attractive molecules is Sonic Hedgehog. If an animal embryo is prevented from making Sonic Hedgehog in the middle of its retina, growth cones from retinal ganglion cells fail to find this point reliably and their navigation becomes chaotic. There is also some evidence of a gradient of repulsion from the edge of the eye.

Once they have left the eye, the first growth cones to do so are faced with a narrow 'corridor', walled with cells that make yet another repulsive molecule.[12] Hemmed in by these repulsive walls, the growth cones can move only one way: along a narrow path to the approximate centre of the developing brain. As thousands of growth cones stream out of the developing retina, laying their axon out behind them, they form the 'cable' of axons that is the optic nerve.

The optic nerves from both eyes converge at the same place in the middle of the brain, and here they are faced with a choice of routes. Some will cross the midline and continue towards the superior colliculus on the opposite side of the brain, while some will swerve and head for the superior colliculus on their own side (Figure 69). The ultimate, functional, reason for this choice of route is to do with the way that we see. The eyes of many animals, especially animals that are hunted by others, are placed on the side of the head. This has the

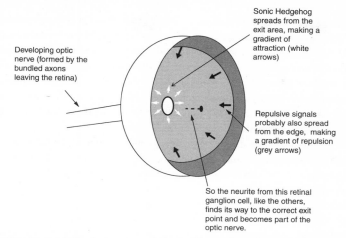

Developing optic
nerve (formed by the
bundled axons
leaving the retina)

Sonic Hedgehog
spreads from the
exit area, making a
gradient of
attraction (white
arrows)

Repulsive signals
probably also spread
from the edge, making
a gradient of repulsion
(grey arrows)

So the neurite from this retinal
ganglion cell, like the others,
finds its way to the correct exit
point and becomes part of the
optic nerve.

FIGURE 68 Pathfinding in the retina at the back of the eye, depicted here as a cup-like structure (which is a reasonable approximation of its real shape). Axons from retinal ganglion cells are guided to the exit point that leads into the optic nerve by attractive cues such as Sonic Hedgehog made in that area, and probably also by repulsive cues from the edge of the retina.

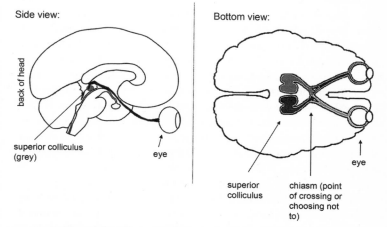

Side view:

back of head

superior colliculus
(grey)

eye

Bottom view:

superior
colliculus

chiasm (point
of crossing or
choosing not
to)

eye

FIGURE 69 The path from the retina to the superior colliculus, in which the visual field is represented in the brain. The left panel shows a side view of the brain, to show the location of the superior colliculus. The right panel, which exaggerates the sizes of the nerves and colliculi a little for clarity, shows how axons arrive at the optic chiasm, and decide whether to cross or to remain on the same side.

advantage that the parts of the land and sky that are viewed by each eye overlap as little as possible, allowing the animal to see almost all the way round them at once and, with luck, to spot a hunter while there is still time to run or hide. With no overlap between their visual fields, the information coming from the two eyes can be processed separately and, in such animals, the optic nerves simply cross in the middle of the brain and each goes exclusively to the structures on the opposite side. We humans, on the other hand, have our eyes looking forward from the front of our faces, a position typical of animals that have to hunt or animals that have to judge distances accurately, for example because they swing from branch to branch in trees.

In humans, more than half of the visual field is seen by both eyes and, because the eyes are some distance apart on the side of the face, this allows the brain to see in three dimensions. To illustrate how this works, hold your finger a foot or two (30–60 cms) in front of your face and close your left eye. Now move your finger so that it is in line with a distant object. Without moving your finger, close your right eye and open your left eye: the finger will have seemed to move with respect to the distant object. The difference in apparent position of objects seen by the two eyes allows the brain to calculate distances very accurately, but to do this the same area of brain needs to get information from both eyes at once. For this to happen, a significant number of growth cones have to choose not to cross the midline but to turn back to their own side of the brain, joining ones coming from the opposite eye. The turning back of some growth cones at the midline is again done by repulsive molecules made by cells already at the midline.[13] Some growth cones detect and respect this repulsion while others ignore it, because they bear different combinations of receptors for these molecules.[14] Beyond the crossing point, whether growth cones actually crossed or swerved back to their own side, the path to the superior colliculus is again defined by repulsive cues that stop growth cones straying off the path. There may well be attractive molecules spreading from the target as well:[15] this is known to be the case for other routes of growth cone migration within the brain, such as the path that connects sensory information into the cerebral cortex.

It is all very well to list the guidance cues that a particular set of growth cones uses to navigate, but this begs the question of how the guidance cues come to be made in such an intricate pattern in the first place. The answer—what little we yet understand of it—mirrors a process that has already been described in the context of the embryo as a whole. As cells in the central nervous system develop, a combination of cues provided by neighbouring tissues and the proteins already present in the cells determine which of their genes will be switched on or off. Some of these genes specify the production of signalling molecules that act as cues for other neighbouring cells and can affect their gene expression. In this way, an initially simple and homogeneous system can organize itself to become very complicated and heterogeneous. Within the nervous system, there is the added complication that, once neurons start to send out growth cones and these growth cones use cues from surrounding cells to navigate and to lay out axons behind them, the axons themselves can act as cues that can either change gene expression in nearby cells or direct the migration of subsequent growth cones from other cells. Like the embryo as a whole, the developing nervous system is a self-creating landscape that adds complexity on complexity, and in which present geography depends on past history.

This method of development, in which the response to one change becomes the stimulus for the next, is a powerful method of increasing complexity, but it entails serious risks. Where small differences made by one mechanism are amplified many-fold by the way that subsequent events build on that initial difference, there is little room for error, and the failure of one system early on can have disproportionately devastating effects later. It is perhaps for this reason that there are so many genetic diseases that have serious effects on brain function.

One example of a brain abnormality that stems from a problem early in neural development is lissencephaly. A variety of mutations can interfere with the normal movement of neural cell bodies as the neural tube thickens, and this means that the layers are not formed properly, the neurons instead making a tube with too little surface area (so little that the brain does not fold to make its usual furrowed surface; 'lissencephaly' means 'smooth brain').[16] Without the normal layers, the brain does not work properly, and children born with the most severe forms of the condition show very poor intellectual development, typically failing to progress beyond the abilities of a child in the first few months of life. They also suffer serious muscle spasms and seizures, and often die very young of a failure to control breathing properly.

Later, various mutations can cause specific defects in the guidance of growth cones, causing both failures of normal connection and the making of inappropriate connections. L1CAM is a cell adhesion molecule that is expressed on certain cells of the brain. Growth cones with the correct receptors to recognize L1CAM will adhere to these cells and be able to migrate along them. Mutations that affect L1CAM function remove this essential guidance cue,[17, 18] and people with such mutations have missing connections between the left and right halves of their brains, and between their brains and spinal cords. As a result, they have problems with movement and other aspects of brain function. Some humans also show mutations in genes that encode proteins involved in the ROBO/Slit system that controls midline crossing. In these people, growth cones that are supposed to cross the mid-line cannot, because they are abnormally sensitive to ROBO. As a result, they suffer visual impairments and problems coordinating movement.[19]

This chapter has used just a few examples of growth cone guidance, with the names of half a dozen molecules, to illustrate what are thought to be general principles that govern the initial wiring of the nervous system as a whole. There is a danger that it might give two misleading impressions. The first is that all wiring is controlled by just two or three molecules. That is definitely untrue: the nervous system as a whole contains many different proteins that guide growth cones, different growth cones being responsive to different ones. These proteins usually work in combinations; even if only two were found together in one place, a choice of any two out of a thousand molecules would give a million possible combinations. In fact, tens of different proteins can be made together in the same place, yielding numbers of possible combinations that have more noughts than would fit on a line on this page. It is a little like the way that the use of just ten digits (0–9) in different combinations to specify uniquely the hundreds of millions of telephones that now exist in the world. Combinatorial use of growth cone guidance molecules is almost certainly the way that the developing nervous system can specify so many different types of connection. The second impression that must be avoided is that we already understand the development of the nervous system well. Nobody does. We have a few partially understood examples from which to learn, such as the ones described in this chapter, and these are probably enough to be reasonably certain that the general principles of how growth cones are guided are understood. Those general principles, though, are a long way from a precise, detailed understanding of how every part of the brain is wired up. There is still very much to learn.

The mechanisms of growth cone navigation described in this chapter set up a basic pattern of connections, but this lacks the precision of the fully formed nervous system. Much of this precision is achieved by using the signals that travel along the axons of a wired-up nervous system to refine the pattern and strength of connections, in a process that continues throughout life. Because it happens later in development, mainly after birth, the mechanisms of this refinement will be discussed later, in Chapter 15.

PART III

REFINEMENT

14

DYING TO BE HUMAN

In the myddest of lyfe we be in death.
Book of Common Prayer

One of the many ironies of human life is that it depends on death—the death of vast numbers of individual cells in an otherwise healthy body. This is not the accidental death of scrapes and injuries that takes place in everyday wear and tear, nor is it the result of cellular assassination by bacteria or viruses involved in infectious disease. Instead, it is a quite deliberate cell suicide, in which cells activate proteins that result in their controlled destruction. It is estimated that more than half of the cells made by a foetus eliminate themselves from the body as part of entirely normal development. Because cells 'choose' to destroy themselves, the process is called 'elective cell death'.*

One reason that cells may choose to die is that they form a tissue that is needed for construction of the body without actually being included in the final product. Tissues like this can be compared to the scaffolding that builders use to make an arched bridge, which is removed once the bridge is complete and self-supporting. The temporary kidneys, mentioned in Chapters 10 and 12, are arguably examples of such tissues.[1] In 'lower' animals, such as fish, this type of kidney remains in place and functions in the adult, but in adult mammals excretion is done by the 'new' type of kidney that was described in Chapter 12. Like fish, mammals make their first blood cells and blood vessels in a complex of

---

* There are several different types of 'cell suicide' (eg. apoptosis, autophagy…) and I am therefore following the same practice used in my book *Mechanisms of Morphogenesis*, and use 'elective cell death' as an umbrella term.

tissues that includes the primitive kidneys. Males also use parts of the drainage system of the temporary kidney for reproductive plumbing. Mammals therefore have to make the structure in embryonic life, even though their adult bodies have no need of it for its original purpose of excretion.

One very visible place in which elective cell death is important is in the hands and feet.[2] These begin as paddle-shaped objects, in which finger bones form by condensation of mesenchyme, all sharing the same mitten-like covering of ectoderm (Chapter 11). In a normal human, the cells that lie between the fingers kill themselves, causing the overlying skin to shrink back towards the palm and to take on the form of a glove rather than a mitten.[3] The importance of elective cell death in the making of fingers has been illustrated by experiments on bird embryos. Chick embryos show a great deal of elective cell death between the forming toes, and they end up with long, well-separated digits that are prefect for scratching in the soil. Duck embryos, on the other hand, show very little cell death between the toes and they end up with webbed feet in which toes are still connected by a tough double-layer of skin and connective tissue, perfectly adapted for swimming. If chick embryos are treated with a drug that prevents elective cell death, they make webbed feet (Figure 70).[4]

As well as being a means to eliminate scaffold tissues, elective cell death is a common feature of permanent tissues, where it is used to eliminate excess cells. Many developing tissues over-produce cells initially and then use signals from other tissues to determine how many, and which, should be kept.

One of the most dramatic examples of over-production of cells can be seen in the motor neurons of the developing spinal cord. Motor neurons are the neurons that connect to the muscles of the body wall and the limbs and are responsible for the movement of the arms, legs, and trunk. The example of the arms will serve to illustrate a principle that applies throughout the body. In normal development, the region of the spinal cord that serves the arms, for example, produces far more motor neurons than will be needed in the adult.[5, 6] These motor neurons send axons into the developing limb and these aim to make contact with fibres of the developing muscles. Some time later, a wave of elective cell death occurs, significantly reducing the population of the motor neurons.

The first insight into what controls the elective cell death came from experiments[7, 8, 9] in which one forelimb (wing) bud of a chick embryo was removed, the other being left in place as a control. The initial large population of motor neurons was produced normally in both sides of the spinal cord. When the

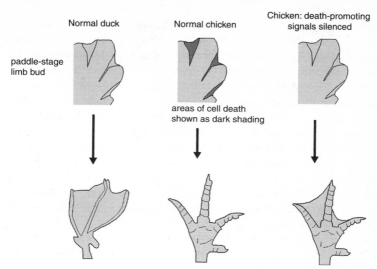

FIGURE 70 The importance of cell death in eliminating webs between fingers and toes. The left panel shows the development of a normal duck foot, in which there is little elective cell death between the toes in the limb bud, and inter-digital webs persist. The middle panel shows the development of a normal chicken foot which, like a human hand, shows significant elective cell death between the digits and produces a non-webbed limb. The final panel shows the result of an experiment in which the signals that normally elicit elective cell death in the chick foot have been blocked by an experimenter; the result is a duck-like foot, suggesting that elective cell death is essential for clearing the webs ('suggesting', not 'proving', because it is not possible to be certain that the method used to block death-promoting signals did not also have some other effect on some yet-to-be-discovered process in the limb: the possibility of 'unknown unknowns', to use Donald Rumsfeld's famous phrase, is a persistent problem in interpreting biological experiments).

wave of cell death occurred, the side of the spinal cord still adjacent to a developing limb showed the normal, modest loss of motor neurons, but the side of the spinal cord against which there was no arm showed a massive loss. This classical experiment suggested that a motor neuron's decision between survival or death may depend on whether its axon succeeded in finding a target muscle to which to connect. The tentative conclusion was supported further by the observation that grafting an *extra* arm on to one side of the body greatly *reduced* the amount of motor neuron death that took place on that side, as would be expected with two limbs' worth of muscles to innervate.

The observations that changing the amount of target muscle can alter the proportion of motor neurons that die, and that normal development involves some motor neuron death, suggest that even in normal development there is not quite enough muscle for all of the motor neurons that are produced. Painstaking biochemical analysis of the developing muscles has revealed that they produce *very limited* amounts of neural survival factors.[10] Initially, motor neurons can survive without these but, as they mature, they become dependent on an adequate supply. Within the cells, a suicide-promoting pathway is already present, and it is only held off by signals from the survival factors. In the normal foetus, the arm muscles do not produce enough survival factors to keep all of the motor neurons alive, and only cells that have made the best connections to the muscles will receive enough. Other cells will have insufficient signal to hold the suicide pathway off and they eliminate themselves. Cells in the over-large population therefore compete with each other for survival factors, and those that have managed to make the best connection with the source of these factors survive.

Over-production of cells followed by selection of the best connected is reminiscent of the process of over-production of young animals, followed by survival of the fittest, that lies at the core of Darwinian evolution.† The parallel is inexact, but both systems provide a means of selecting for the 'best' in the face of random variation. In Darwinian evolution, the variation is in the combination of genes possessed by each competing young animal. In the embryo, the variation is mainly in the accuracy with which axons find their way to the target. In both cases, the individuals that happen to be best placed to survive do so, and the others die. By using such a system, the embryo reduces the need for super-accurate pathfinding by axons and becomes much more tolerant of errors.

The principle of cells competing for target-derived survival signals is by no means limited to the motor neurons of the spinal cord. It is also seen in the sensory system and in many areas deep within the brain itself. Importantly, it is also seen in non-neuronal organs. Indeed, it is so general that it has given rise to the 'trophic hypothesis';[11] this postulates that all cells depend for their survival on limited quantities of survival factors released by other cells. One of the scientists who developed this hypothesis was so confident of its truth, when he

---

† A graduate student once suggested to me that it is also reminiscent of universities training far more scientists than there are research positions to fill. This may be true: in education, as in the embryonic nervous system, it is difficult to be sure at the start which candidates will develop exactly on target and which will end up blundering about totally lost in an environment that is not right for them. In both cases, overproduction followed by selection works for the system as a whole, but does not go well for those not selected. It will come as no surprise that the student who saw this parallel was one of the successful ones, and is now doing inspiring work on links between development and cancer.

first described it in talks, that he used to offer a substantial prize to anyone who found a counter-example (abnormal cells and very early embryos excluded). Nobody succeeded in claiming it.

The trophic hypothesis is useful to human development at both very short and very long timescales. At short timescales, it means that cells that have formed in the wrong place, or have wandered away from where they are meant to be, will be far from their intended interaction partners and will simply kill themselves. They will therefore not cause problems. At much longer timescales, it is one of the things that makes evolution of complex bodies possible. Consider two hypothetical animals, of the same body shape, one of which produces exactly the right number of motor neurons to serve its arm muscles in the first place, and the other of which over-produces them initially and then culls the excess, as we actually do. Now imagine that the environment of the animals changes, and a new ecological niche opens up that can be exploited by animals with strong arms: examples might be lifestyles that involve digging or that entail swinging through trees. For an animal of the first type to acquire larger but still functional arms, there would have to be two types of mutation brought together in the same individual, one that makes the arms larger and the other that makes the pool of motor neurons larger, by just the right amount. For an animal of the second type, though, just the mutation for larger arms would be enough, because the number of motor neurons that survive would adjust automatically for the new, larger source of survival factors. Getting just the one mutation would be a low probability event; having two in the same individual would be a *very* low probability event and would probably entail a long wait. Animals that use overproduction of cells followed by target-dependent survival are therefore better placed to win evolutionary races—to evolve faster and thus be the first to exploit new environments. It is probably not at all surprising, therefore, that complex animals like mice and humans, that have a long history of evolutionary innovation behind them, are built on these principles: if they were not, they may never have had time to evolve in the lifetime of the Earth.

The dependence of cells on survival signals from other cells has interesting clinical implications, and physicians' growing ability to manipulate signals for survival of cells is already helping the survival of people. Cancer cells, more or less by definition, have lost their normal growth control mechanisms, but many retain their dependency on survival signals. This is a potential Achilles' heel for certain tumours, because it allows the possibility of persuading the cancer cells to kill themselves in a way that is much less damaging than the broadly toxic approach of conventional chemotherapy. The tumours that are most vulnerable to this new approach are those of inessential tissues, because it will not be a life-threatening disaster if the treatment persuades some normal body cells of the same tissue to kill themselves too. Survival of cells in the normal prostate gland is dependent on testosterone signalling; indeed the observation made by Karl Vogt in the 1840s, that the prostates of young male cattle shrink dramatically following castration, was the first clear description of elective cell death following the failure of a survival signal. Many prostate cancers and pre-cancerous growths retain their dependence on testosterone and other hormones, and drugs that block testosterone signalling can be very valuable in reducing the tumours.[12, 13] Similarly, many breast cancers depend on oestrogen as a survival signal. Tamoxifen is a powerful inhibitor of oestrogen signalling in breast tissue and it has been a very effective drug against breast tumours of many patients.[14] Using this approach against tumours of essential tissues has the complication that completely blocking an essential survival signal risks destroying normal tissue as well as the tumour. The messy arrangement of tumours does bring many of their cells rather far from sources of survival signals, so it may still be possible to find drug doses that allow just enough signalling to persist to support normal tissue, which will be in exactly the right place, but not enough to support survival in a chaotic tumour, most of whose cells will be quite far from their source of support. Coupled with other anti-cancer therapies, manipulation of survival signals may become valuable against a greater range of cancers in the future.

15

MAKING YOUR MIND UP

Systems—systems—systems—you can't escape them because nature is systematic, and man is a natural phenomenon, and his intelligence is a natural phenomenon.

Donald Crowhurst

Of all the things that a growing human has to develop, the most remarkable and the most difficult to understand is the mind. We still know very little about how the higher functions of the mind work although it is clear, from the specific and predictable effects of different types of brain damage, that the mental phenomenon of 'mind' emerges from neural activity in the physical brain.

Like the spinal cord, the brain develops from the neural tube. The head end of this tube first develops three bulges, the future fore-brain, mid-brain, and hind-brain (Figure 71). The growth of these bulges is associated with vigorous cell multiplication in the walls of the neural tube (Chapter 13), but it may also be driven by internal pressure in the central fluid-filled cavity of the brain, the drain for which is blocked temporarily at this stage of development. Once the three basic bulges have formed, the sides of the fore-brain billow out further to form the two sides of what will be a new structure, the end-brain, and the other bulges also begin to sub-divide themselves into distinct regions. Again, the process of development follows the sequence we have seen many times before, of an area dividing itself up into zones, growing, dividing the zones into sub-zones, growing, and so on.

In a simple vertebrate, such as a fish, the anatomy of the brain remains as a straight tube with swellings and is easy to relate to the embryo: that is why

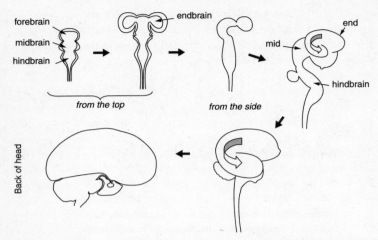

FIGURE 71 The basic shape of the human brain develops by a sequence of bulging and folding events.

dissection of the dogfish was such a popular exercise in school biology before dissection became unfashionable. In mammals, and especially in humans, the need to cram a lot of brain into a rather small and complicated head is satisfied by bending and folding the basic structure of the tube until it becomes rather hard to recognize. To add to the complications, human end-brains grow massively and creep around the rest of the brain until they cover it (Figure 71), and then go on to become furrowed to cram even more surface area into the available space.

As far as we know, the structures of the brain are all produced by the same basic developmental mechanisms that produce the simpler spinal cord: there is cell division to create layers of cells (many, in the case of the brain), cell signalling to tell cells what to be and to refine an initially simple pattern, folding and re-folding of the tube, navigation by growth cones to lay down an accurate 'wiring' pattern (Chapter 13), and death of surplus cells (Chapter 14). The difference is that there is much more of these things going on, particularly in advanced vertebrates such as ourselves whose brains have come a long way from being a tube with a few bulges. Many of the signals involved in the wiring and the life-and-death decisions have now been identified, but to recount them would be tedious and the mass of detail would not bring with it any new principles, just baroque variations on themes we have already explored. More interesting is the question of how this mass of anatomical detail allows the development of the epiphenomenon of the brain—a responding, learning, thinking mind. Neurobiology still understands far too little about how the mind works to attempt anything like a full explanation. It is possible, though, to take one element of the higher functions of mind—learning—and to provide a tentative sketch of how the cells of a developing foetus build structures that make learning possible.

To the best of our current knowledge, learning works by altering the connections between neurons in the brain in the light of the signals they carry. Connections between neurons are, as we have seen earlier, called synapses: in each synapse, a bulb on the end of an axon of one neuron, the sending neuron, lies extremely close to the membrane of the receiving neuron. In some types of synapse, electrical signals pass directly from neuron to neuron. In other types of synapse, electrical firing in the axon of the sending neuron causes it to release chemicals (neurotransmitters) into the tiny space between it and the receiving neuron, and these drift across the gap and bind to receptors on the receiving neuron, stimulating it or repressing it depending on the types of neurotransmitter and receptor involved. Most neurons in the brain receive synapses from thousands of others. Since there are tens of billions of neurons in the brain, and up to one thousand synapses per neuron in an adult, there are literally trillions of synapses. In general, a signal from one synapse is not in itself strong enough to make the receiving neuron fire, or at least not to fire much, but simultaneous signals from several synapses can add up to enough of a signal to fire the receiving neuron. This means that neurons fire in response to an integrated combination of signals they receive, which would in principle allow different types of information (sight, sound, memory, desire...) to be brought together to determine an action. We know this is what happens in very simple animal models, and for the moment we just assume it scales up to a full-blown human brain (there is no reason to suppose it does not, but it would be arrogant in the extreme not to assume that continued study of the brain will reveal some surprises). Most receiving neurons send axons on to other neurons, to create complex neural 'circuits'. Some neurons, called motor neurons, send axons directly to muscles so that firing of the neuron causes muscle contraction. This allows thought to be turned into action.

There are two main ways of altering the connections between neurons. One is to keep the connections anatomically the same but to alter, biochemically, the efficiency with which each transmits a signal. This has the merit of being fast. The other is to alter the wiring pattern itself by destroying existing connections and growing new ones. This must be slower, because of the time needed to grow new axons (growth cones travel at about one neuron diameter an hour). About sixty years ago, the Canadian neuroscientist Donald Hebb proposed a mechanism by which the first type of learning, altering the strengths of existing connections, might occur automatically.[1] Hebbs' mechanism is easier to discuss if we have in mind a specific, simple type of learning; the conditioned, or Pavlovian, reflex.

The Russian physiologist Ivan Pavlov discovered 'Pavlovian conditioning' while working on the physiology of the digestive system. Dogs naturally salivate when they see or smell that they are to be are presented with food, because activity of the salivary glands is controlled party by the brain and not just by mechanical presence of food in the mouth. In his experiments, Pavlov presented dogs with an irrelevant stimulus, such as the sound of a whistle or a bell or the sensation of a mild electric shock, just before giving them their food. He used the same irrelevant stimulus consistently for each dog, and did this for several mealtimes. After this period of conditioning, he found that presenting the dogs with the irrelevant stimulus alone, with no food, would make them salivate. They had learned and remembered that the stimulus was a sign that food would arrive. Dogs have very complex mammalian nervous systems but the same kind of conditioning can be found in simpler vertebrates; the tropical fish *Chromobotia macrocanthus*, for example, quickly learns to associate the sound of a shaken food tub with feeding and go into an anticipatory frenzy of activity when the sound is heard. This learning can even be demonstrated in fruitflies, whose nervous systems are very much simpler than those of dogs. If fruitflies that have received no conditioning are placed in a tube between two chambers, each of which has its own distinctive odour, the flies explore the chambers randomly and show no preference. If they are first exposed to one of the odours alone, this makes no difference to their subsequent behaviour. If, though, they are first exposed to one of the odours while at the same time being given a series of electric shocks, and are then placed in the tube, they will choose to avoid that odour and go to the other one instead. Clearly, they learn to associate the odour with the shocks.[2, 3]

Hebb proposed that receiving cells can biochemically alter the efficiencies with which they respond to the synaptic signals they receive, on a synapse-by-

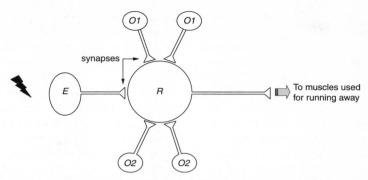

FIGURE 72 A simple model of conditioned learning in fruitflies. Neuron R receives initially weak inputs from neurons O1 reporting odour 1 and neurons O2 reporting odour 2, and also inputs from neurons labelled E that are stimulated strongly by electric shocks. In reality, greater numbers of neurons of each type are involved, but only a few are drawn here for clarity.

synapse basis according to a specific rule: the efficiency of response to a particular synapse is increased if that synapse fires when the receiving cell itself is firing. To illustrate this, consider a simple model of a part of the nervous system of the fruitflies subjected to the odour and shock experiment. The model, illustrated in Figure 72, has four types of neuron (an unrealistic oversimplification of real fly brains, but helpful for explaining the basic principle). Type O1 neurons are activated by the presence of odour 1, and type O2 neurons are activated by the presence of odour 2. Type E neurons are activated by electric shock, and Neuron R, to which all connect will, if activated, evoke a 'run away' response in the animal. Before the learning has taken place, the connections of both type O1 and type O2 neurons to R are too weak to activate it, so neither odour causes the animal to run away. The connection of type E neurons is strong enough, though, to activate R when the type E neurons are stimulated by an electric shock. If, when the shock is applied and R is activated by signals from type E neurons, odour 1 is present, type O1 neurons will be active. Hebb's condition will therefore be satisfied and the response of R to O1 synapses will be strengthened. If this happens often enough, the response will be so strong that firing of O1 neurons will be enough to activate R even with no help from type E neurons, and the fly will have learned to run away from odour 1 alone without waiting for a shock. The response to type O2 neurons, will not have been strengthened by any of this, so odour 2 will not cause the fly to run.

In the decades since Hebb published his idea, some of the biochemical mechanisms underlying it have been discovered. In many of the systems studied, including the fruitfly, the synapses involved use a neurotransmitter called glutamate. The receiving cells have two types of receptor for this neurotransmitter. One type, AMPAR, is straightforward in its action: each molecule of AMPAR that has bound some glutamate activates a complex of proteins inside the receiving cell and makes a modest contribution to trying to make that cell fire. If enough AMPAR molecules bind glutamate, and if the sensitivity of the complex inside the cell is high enough, then the cell will fire. The other type of receptor, NMDAR, is far from straightforward because it behaves differently according to whether the cell bearing it is already firing or not. In a resting neuron, the NMDAR molecules cannot do anything even if there is plenty of glutamate about. In a firing neuron, though (that is, a neuron that is receiving enough total input from all of its synapses to make it fire) the NMDAR molecules are able to respond to glutamate and to send their own signals into the cell. These signals do not activate the cell much directly, but rather they alter the AMPAR system local to that synapse to enhance the amount of signal that a given amount of glutamate can trigger through it. The NMDAR receptors are therefore at the heart of Hebb's system, because they are active only when the receiving cell is already active *and* the particular synapse is active (presence of glutamate), and they can alter the strength of the synaptic connection by making its AMPAR system more sensitive (Figure 73).

The importance of the NMDAR system to learning has been demonstrated very clearly in the fruitfly odour and shock experiment. Genetically engineered fruitflies have been created that allow experimenters to switch off production of NMDAR proteins at will, and switch them on again later. Because NMDAR protein molecules are relatively short lived, the flies have almost none left by fifteen hours after the switch off. Under these conditions, they are very defective in their ability to learn to associate an odour with a shock. Once production of NMDAR has been switched on again, they are able to learn normally.

Hebbian adjustment of the strengths of synaptic links provides one mechanism for learning, but not the only one. There is good evidence, in humans as well as in lower organisms, that the very wiring patterns of the brain can alter according to experience. One of the best-studied examples of this is again provided by the development of the visual system, and it follows on from the chemically guided coarse mapping of eye to brain that was described in Chapter 13. This chemically guided connection of retinal axons to neurons in the brain is

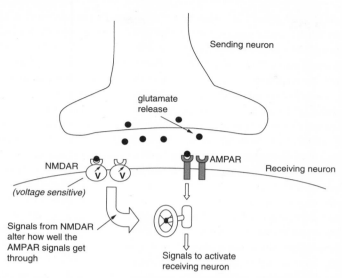

FIGURE 73 The action of the voltage-sensitive receptor, NMDAR, in controlling how well AMPAR receptors can stimulate the receiving cell. The valve symbol represents networks of signalling proteins and systems that control how much AMPAR is in the membrane, that together regulate how much signal a given amount of glutamate can trigger from the AMPAR system.

only approximate. As they make synaptic connections in the superior colliculus and in other areas of the brain, branches of the axons from a particular part of the eye will connect to a lot of correct cells, also connected to by their neighbours, but will also connect to a few wrong cells too. Left in this condition, the system would result in blurred vision that would not do justice to the capabilities of the eye. The wiring pattern is therefore refined, after birth when the eyes are open, by activity-dependent remodelling of connections. The mechanism for this is similar to Hebb's, but this time not only the strength of synaptic connections but also their very survival depends on a synapse's own firing coinciding with the firing of the receiving cell. If a synapse from an axon from, say, the top left of the eye connects to a neuron to which axons from other cells in the top left of the eye also connect, then it will be firing when they do, because all will be reporting the same events in the visual field. The combined action of all of these synapses will fire the receiving neuron, and all of these synaptic connections will be strengthened. As a receiving cell fires more often, it becomes less willing to support weak synapses whose firing never coincides with that of the receiving cell. A synapse whose own firing does not coincide with the firing of the receiving cell must be responding to different stimuli from the majority of synapses on the receiving cell, which means it must come from a different part of the eye or must be sensitive to some other aspect of the visual field: either way, it does not belong. By disengaging from such out-of-synch synapses, the receiving cell frees itself from inappropriate connections. The freed axon end can then try its luck connecting to a different cell, and repeat the process (still guided by the basic chemical cues described in Chapter 13) until it can settle on a neuron that is being stimulated by other synapses with the same firing pattern.

Activity-dependent remodelling of neuronal connections has an interesting consequence for animals such as humans and cats, whose eyes look forwards at the same scene to produce binocular vision. For such animals, axons that carry signals about some part of the visual field from one eye map to approximately the same part of the visual cortex as axons from the other eye that represent the same part of the visual field. If one eye of a newborn animal is covered for a few weeks, its axons will be electrically quiet because there will be no image to report. The other eye will still work normally and the synaptic connections of its axons will be stabilized, while those from the shut eye will be preferentially lost. When the temporarily closed eye is uncovered, it is still anatomically perfect but it no longer communicates with the visual cortex and is effectively blind. This is the condition seen in humans with ambylopia, or 'lazy eye', in which one eye is not properly 'wired' to the brain. In both humans and animals, the condition can be corrected by shutting off activity from the dominant eye, for example by fitting a child with an adhesive eye patch for a few months, to give the subordinate eye a chance to establish connections of its own. The two eyes then need a period of working together before their maps fully coincide.

The visual map is not the only spatial map in the brain: sound is mapped too, so that we can locate the source of a sound in space. Humans are not particularly good at this, but owls, which hunt in conditions of low light, can locate sources of sound very well using the slight difference in volume and timing of a sound as reported by their two ears. In owl brains, processed input from the ears and from the eyes is brought together in the tectum, the bird equivalent of the superior colliculus described in Chapter 13, so that each part of the tectum responds to integrated light and sound signals from a particular direction in space.[4] Alignment of the visual and auditory maps depends on activity-dependent remodelling, and this has been demonstrated by a famous experiment that involved owls wearing spectacles. The spectacles were made with prisms that shifted the visual field sideways so that, when the owl's eyes pointed straight ahead, they actually saw an image that corresponded to looking a few degrees to the right. Within the tectum, the connections slowly rearranged themselves so that the unaffected input from the ears again aligned correctly with the displaced input from the eyes. The process of re-alignment took around two weeks and was stable as long as the owl continued to wear the spectacles. Removing the spectacles while the owl was still young, less than about six months old, resulted in its reorganizing its connections yet again so that the auditory and visual maps again coincided. Beyond about six months, the owl's brain is much less able to undergo major re-organizations:[5] this is perhaps similar to the way that young humans can learn new languages easily while adults struggle.

The ability of the brain to modify its connections according to activity provides it with a mechanism for hard-wiring the association between signals and bringing together inputs that coincide in the external world, so that they control the same neurons. Neurons that fire together, wire together.

Hebbian learning to establish associations between different neural signals, and activity-dependent remodelling to hard-wire the associations, have obvious roles to play in improving the processing of sensory information and for promoting very basic types of learning, such as the Pavlovian conditioned reflex. These basic elements of brain function seem a long way from the 'higher' functions of mind, such as debating economics or analysing neural systems or writing love poetry. Can even these higher functions be built from such simple cellular mechanisms? We do not know for sure, but there is at least an argument for their playing a major role. Most of the highest functions of mind rely on associations, between objects, places, ideas, memories, and a host of other things, and the Hebbian system and its relatives are a powerful mechanism for associative learning.

One example of a mental faculty that relies heavily on association is our facility for language. The essence of most language is that the spoken sound or the sequence of written symbols used to signify something is arbitrary, and bears no relation to the nature of that thing. A few words, typically those used to describe animal sounds, are onomatopoeic (*quack, moo*), but most are not. Nevertheless, a rose by any other name—*rhosyn, ruusu, роза, tăng,* 玫 瑰—does smell as sweet. The understanding of language, therefore, depends in part on forming stable neural associations between actual objects, locations, etc., and words. In principle, this is similar to dogs forming stable neural associations between ringing bells and the arrival of food, and may use similar mechanisms of synapse stabilization and synapse elimination. Even beyond language, much of our ordinary, everyday understanding of the world is based on associations, between a face and a name and a remembered kindness, between a geographical location and the opportunity to buy bunches of sweet-smelling roses, between a movement of the right foot and a car stopping safely. Some recent studies have shown clear associations between areas of the brain known to be important in associative learning and lifestyles that require the recall of vast amounts of geographical information, for example driving a London taxi-cab. They provide another example of the brain altering itself in response to the environment, this time in relation to higher functions rather than the lower-level functions of seeing and hearing.

There may well be much more to mind, and consciousness than Hebbian synapses and activity-dependent remodelling. Nevertheless, these aspects of neural development, which we do understand, more-or-less, seem to be a critical prerequisite for building a mind. They provide yet another illustration of the way in which the construction of the body depends on cells organizing their relationships according to the signals that they receive. The genes that are active during the building of a brain do not work by specifying an exact final anatomy, for that would be self-defeating in a system whose structure has to reflect what it has learned. Rather, the proteins specified by those genes together make mechanisms that strengthen, weaken, or destroy connections according to how well signals upstream and downstream of those connections coincide. Constantly comparing outputs with inputs, these systems refine the brain, changing the way that signals associate and noting what has been learned in a three-dimensional script of changing neural connections.

These discoveries, some of which are now decades old, have a great deal to contribute to the tiresome nature/nurture debates that continue to drag on between enthusiasts for genetics and enthusiasts for sociology, in which one side views most mental attributes as being determined genetically and the other assumes most to be determined by the environment. The emerging truth is that, for almost everything that matters, genes and environment have to work together: the action of the proteins specified by genes builds neural machines that can decide how to connect according to the environment in which they are located. Deleterious genes or a 'bad' environment both result in mental deficiencies. 'Good' genes create a brain with the potential to make a healthy mind, but that potential will be realized only if the environment of childhood provides the stimuli to make the right internal connections. For social animals like humans, that means more than the simplicities of visual experience and sound: it means the richness of language, of interaction, of play and of love. We now know—we don't just guess from particular political standpoints, but actually *know*, from cold, factual images in MRI scans—that the brains of children subjected to frequent and severe verbal aggression and abuse grow to differ physically from those of children growing in a more nurturing environment.[6, 7] The same is true, but in different areas of brain in each case, for children who suffer physical violence, including frequent corporal punishment,[8] or sexual abuse.[9] The images give a haunting illustration of the words of novelist Beryl Bainbridge: *Everything else you grow out of, but you never recover from childhood.*

16

A SENSE OF PROPORTION

You've got to do your own growing, no matter how tall your grandfather was.
Irish Proverb

Just beyond the end of a bridge over Venice's Grand Canal, in the exhibition galleries of that city's Academy of Art, lies one of the most iconic images of the Renaissance. Drawn in ink by Leonardo da Vinci, it depicts a male figure with his legs drawn both together and apart, and his arms drawn both outstretched horizontally and reaching slightly up to the level of the top of the head. Around the body are a circle centred on the navel and extending to the soles of the feet, and a square the height of the man (Figure 74).

The text that accompanies the drawing, written in the mirror script characteristic of Leonardo's secret notes, lists a number of facts about the relative sizes of the body parts. These include rules such as the span of the outstretched arms being equal to a man's height (a point made by the square in the drawing), the distance from hairline to chin being a tenth of a man's height, the distance from elbow to the tip of the hand being a quarter of a man's height, the length of a foot being a sixth of the man's height, the length of an ear being a third of the length of a face, and so on. There are thirteen rules in all. They are not original to Leonardo, but come rather from the Roman architect Vitruvius, who set out these rules in the first century BC. In his honour, Leonardo's illustration of these rules is usually called 'Vitruvian Man'. As the modern anatomical artist Susan Dorothea White has pointed out by her pastiche drawing *Sex change for Vitruvian Man*, essentially the same rules also apply to the bodies of women.

FIGURE 74 *Vitruvian Man*, Leonardo da Vinci (Wikimedia Commons) CC gaggio1980-Fotolia.com

Although it is now recognized that these 'rules' of proportion are averages rather than absolutes, and that plenty of people have slightly longer faces or smaller feet or bigger ears than Vitruvius' rules specify, it is still remarkable how closely all but the most abnormal bodies follow this general pattern. How do tiny cells manage to fix, with such accuracy, the shapes and proportions of a body so much larger than themselves?

The proportions that emerge with maturity are not those of the embryo, for relative sizes of body parts alter in a predictable way through foetal life, infancy, childhood, and adolescence. At birth, for example, the head is much larger and the limbs much shorter, in proportion to the body, than they will be in the adult, and the body as a whole is much smaller in absolute terms than it will be later on. Throughout this growth, the various parts of the body control their size so that they are always in the correct proportions to each other. In this way, the remarkable near-symmetry of humans is maintained. The body parts them-selves can be far from in contact with each other (consider the left and right feet) and, even in a newborn baby, the length of large parts such as limbs exceeds the length of an individual cell by ten thousand times. How, then, do body parts measure themselves? What mechanisms give the growing body its sense of proportion? The simple answer is that we do not yet know, but it is possible, by pulling together data from disparate experiments on animals that range from fruitflies to mammals, to make some educated guesses about the processes involved.

The first issue to consider, before addressing problems of proportion, is the control of body size as a whole. Much of what we understand about the control of body size comes from studying individuals in whom it is altered in some way: dwarfs and giants.

It was observed many years ago that gigantism, the production of a body very much larger than average, is often associated with tumours of the pituitary gland. When the tumours are active during the normal period of childhood growth, the result is a very large human, from seven to over twelve feet tall (2.1–3.6 m). Immense as these giants are, their bodies still have normal proportions.

The pituitary gland is a complex organ that secretes many hormones, but the one that is most important in terms of size control is called Growth Hormone. A healthy pituitary gland releases modest amounts of Growth Hormone in a series of pulses, one every few hours, with activity generally being highest during sleep. The average concentration of the hormone is high during early childhood, when the body is growing quickly, and it shows a sharp decline to the modest 'adult' level around the age of 18 to 20. Humans who make very little Growth Hormone are short, often as short as four feet (1.2 m), but proportions of the body remain approximately normal, so that someone so affected would still be able to be a reasonably convin-cing model for Vitruvian man. A relationship between the amount of Growth Hormone and the size of the body does not itself demonstrate which is cause and which effect, but the observation that children with low Growth Hormone can be restored to near-normal growth by injecting them with the hormone[1, 2, 3] is a clear indication that the amount of Growth Hormone determines body size.

Growth Hormone does not affect cell growth and proliferation directly. Rather, it causes some cells, most particularly those of the liver, to make a second long-range signalling molecule called Insulin-Like Growth Factor I

(IGF-I: the related molecule IGF-II controls growth in foetal life).* It is the IGF-I that communicates instructions about size to most cells of the body. The process by which Growth Hormone controls the synthesis of IGF-I involves its binding to a specific receptor and triggering an internal signalling pathway that culminates in the expression of the IGF-I gene.[4] Some people carry mutations in the receptor so that their cells are inadequately sensitive to Growth Hormone.[5] The result is a type of dwarfism called Laron Syndrome. It is characterized by an overall very short stature and undersized internal tissues such as those of the heart; the proportions of the skeleton are, however, normal. (Curiously, patients with Laron syndrome tend to live a long time. Animals, ranging from roundworms to rodents, that make lower-than-normal levels of IGF-I tend to outlive normal animals of their species: the same effect may be at work in these people).[6]

Not all abnormalities of growth control result in normally proportioned individuals. A vivid illustration of one that does not is provided by a second artist, one who lived about four centuries after Leonardo da Vinci painted his Vitruvian man. Henri de Toulouse-Lautrec, the post-impressionist famous for paintings evoking the bohemian decadence of *fin-de-siècle* Paris, had an unusual body shape. His face and trunk were of normal size and in proportion to each other, although his head had deformities due to late and incomplete closure of the gaps between skull bones. He had a normal figure at the age of 13 but then his legs stopped growing, although his trunk continued to grow. By adulthood, his legs therefore looked short so that he stood only 5 feet (1.5m) tall. His bones were brittle and painful.[7] Most, but not quite all, modern clinical geneticists who have re-examined his case conclude that he suffered from the genetic condition now called pycnodysostosis.[8, 9, 10]

Pycnodysostosis, a condition so rare that only about two hundred sufferers have ever been described, is caused by a mutation in the gene encoding an enzyme that, amongst other duties, liberates IGF-I that has become trapped in bone.[11, 12, 13, 14] In the absence of a working enzyme, IGF-I remains trapped, unable to drive growth. Patients with this disease can be treated fairly successfully with doses of extra Growth Hormone designed to keep their circulating IGF-I within set limits.[15] Another type of dwarfism, much more common than

---

* IGF-I and IGF-II are 'insulin-like' in terms of their structure: they and insulin itself probably evolved from a single ancestral gene. They are different enough from insulin, though, to make them incapable of performing insulin's direct sugar-management functions.

pycnodysostosis and affecting around one in twenty-five thousand people, is achondroplasia. This is caused by a mutation in a signal receptor, the mutation disrupting normal growth of limb bones.[16, 17, 18] People with achondroplasia have short limbs in relation to their bodies, and also some modifications of the shapes of limbs and some other body parts.

The body shapes of Toulouse-Lautrec and of people with achondroplasia demonstrate that the inability of some parts of the body to keep up with general growth does not necessarily prevent the rest of the body from carrying on growing regardless. Proportion is not, therefore, maintained by every body part checking that it is not outgrowing every other. Instead, different parts of the body must make their own responses to Growth Hormone, IGF-I, and any other relevant hormones. This is even true for two body parts of the same type. In one particularly illuminating experiment done two decades ago, the growth of one leg of a rabbit was inhibited by the local injection of a drug. The leg on the other side of the rabbit still grew normally, to result in a lop-sided animal. Clearly, the symmetry that normally exists in leg length is not driven by the growing legs communicating and regulating each other's elongation.[19, 20]

There is another important message from the limbs of Toulouse-Lautrec and people with achondroplasia. In both cases, the primary biochemical defect affects specifically the growth of the long bones of the limb. There is no direct effect on the growth of skin, muscle, nerves, blood vessels, etc., yet these tissues did not grow to make some floppy mass composed of a normal amount of these soft tissues, all centred on abnormally short bones. Rather, they all grow to be an appropriate size for the short leg. This illustrates a deep dichotomy in size control: some tissues of the body, like the bones in the rabbit experiment, regulate their own absolute size and may be considered 'master' determinants of body size, while other tissues pay attention, not to their absolute sizes, but rather to their sizes relative to the masters'. Tissues of this second type are in a sense 'slaves' when it comes to size control: their job is to keep up, but never to overshoot, the growth of the master tissues. The problem of size control therefore divides into two sub-problems: first, how do the 'master' tissues measure their own sizes, and second, how do the 'slave' tissues match themselves to their masters?

Let us first consider ways in which the growth of 'master' tissues, such as bones, is controlled, taking limbs as an example. Growth of limb bones does not take place throughout, but rather in a specialized zone called the 'growth

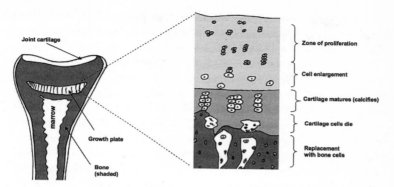

FIGURE 75 The growth plate of a growing limb bone.

plate', near but not at the ends of a bone. Each growth plate is divided into a series of zones (Figure 75). At the outer end of the growth plate, cells proliferate; this proliferation is one of the direct drivers of bone elongation, although not the main one. At the inner border of the zone of proliferation, cells change their behaviour and start to lay down cartilage, a softer precursor of bone. This change effectively moves the inner border of the zone of proliferation one row of cells further out, so that a new row of cells now find themselves at the border. Once they have had time to respond to their new situation, these cells too will switch to making cartilage, and so on. The result of this is that the zone of proliferation moves on, adding cells at its outer end as a result of proliferation itself, and leaving cells behind to make cartilage at its inner end.

In making cartilage, cells take up more space, partly because they become larger and partly because they exude space-filling, jelly-like molecules that separate the cells. This increase in tissue volume is the main driver of bone elongation. As time goes on, the cartilage matures and the cartilage-making cells within it die, to be replaced by bone-making cells from the adjoining mature bone. These cells invade the cartilage and gradually convert it to bone proper. By the time this has happened, the cells in the proliferation zone will have multiplied more, and the whole process will repeat, steadily elongating the bone as it does so, with the zones of the growth plate moving ever outwards. Within this system, the speed at which bone elongates is set mainly by the rate at which cells in the outer part of the growth plate proliferate, and the rate at which cells drop out of proliferation to get on with making cartilage.

From our current knowledge, which is certainly incomplete, the rates of proliferation and of dropping out of proliferation to make cartilage are set by two types of signal: internal signals that organize the growth plate, and external signals that tell it how hard to work. Internal, organizing signals are needed to make sure that enough proliferation takes place to maintain the growth plate against the 'loss' of cells from each of the zones as they mature to become cells of the next zone. The most mature cartilage-making cells, doomed to die and be replaced by bone, secrete a signalling protein that encourages cells at the inner border of the proliferation zone to change behaviour and to commit to making cartilage. The fact that the signal to do this comes from the mature cells themselves means that there is an automatic balance between the number of cartilage-making cells that have matured completely and are heading for death, and the number of cells in the proliferation zone that are instructed to commit to being new cartilage-making cells. This balance maintains the size of the cartilage-making zone as it moves outwards.

The danger of the signalling system described above is that it risks depleting the population of the proliferative zone by making its cells decide to switch to making cartilage faster than they can be replaced by the activity of cells still proliferating. This danger is avoided by another signal that is made by cells that have just stopped proliferating and have switched to making cartilage. This signal spreads out, through the developing bone. Right on the outside edge of the bone, beyond the growth plate itself, specialized cells respond to it by making yet another signalling protein.[21] This spreads back to the proliferative zone of the growth plate, where it causes cells to multiply more rapidly (Figure 76).

Acting together, these two signalling systems, the one saying 'mature!' and the other saying 'proliferate!', ensure that the right number of cells enter the maturation pathway to replace ones moving on to become fully mature and then die, and that the right amount of proliferation always makes good the numbers of cells lost to maturation. The system as a whole therefore remains stable.

The main external regulator of bone growth is the Growth Hormone–IGF-I system already described; for simplicity, this will be referred to simply as 'Growth Hormone' in the rest of this chapter. This seems to alter how much proliferation takes place and therefore joins the internal signals in regulating how many cells are available to start making cartilage. The influence of Growth Hormone is not sufficient to overwhelm the self-organizing capacity of the growth plate, so its structure maintains itself properly, whether the individual

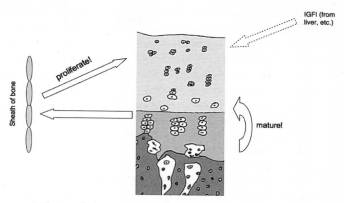

FIGURE 76 The growth plate maintains itself as it grows by a variety of signals, some internal and some proceeding via a relay station in the sheath of the bone.

happens to be fast growing or slow growing. Presumably, the growth plates of different types of bones (finger, thigh, etc.) have different sensitivities to Growth Hormone, so grow at different rates to produce bones with their characteristic relative proportions. Whatever system ensures this, it must be easily mutated, given the great variety in bone proportions in monkeys, apes, and Man.

How, though, do the bones of the same type, such as the thigh bones of the left and right legs, end up an approximately equal size? A hint comes from a variation on the rabbit leg experiment described earlier in this chapter. The variation, which was described in the same report as the basic experiment, was to restrict the growth of one leg so that the rabbit became lop-sided, and then to release the restriction. The result was startling—the short leg started to grow quickly again, more quickly than the normal leg was growing by then, and caught up to become normal length,[19] Clearly, there must have been enough Growth Hormone in the body to support rapid growth, yet the normal leg was not responding to it with the same enthusiasm as the leg that had been released from temporary inhibition. Why should the two legs show different amounts of response to the same concentration of circulating Growth Hormone? Could growth itself make the growth plate become less sensitive to growth-promoting hormones? Such a mechanism, if it existed, would provide an elegant way of having two legs the same length, for any leg that for some reason lagged behind would retain more sensitivity to Growth Hormone and would therefore be able to catch up.[22]

The details of one of the internal signalling pathways suggest one possible mechanism by which the rate of a bone's growth might depend on how much it has already grown. As explained above, the positive effect of maturing cells on increasing cell numbers in the proliferative zone is not direct, but a signalling loop to the outside layer of the growing bone and then back again to the prolif-erative zone.[21] When the bone is small, the outside layer will not be far from even the middle of the growth plate, and the out-and-back signalling loop will be short and efficient. When the bone is larger, the edges of the growth plate may still be close enough to the outer layer of bone to enjoy efficient signalling, but the middle will be further away and receive less of a proliferation signal. On average across the growth plate, the rate of proliferation will fall, and will carry on falling the bigger the bone becomes and the longer the signalling loop there-fore becomes. For a given amount of Growth Hormone, bones will, therefore, become less enthusiastic responders the larger they grow. This would explain the normal symmetry between left and right limbs fully, and would also explain the data from the rabbit experiment. It must be stressed, though, that the ex-planation in this paragraph is just a speculative attempt to draw various discov-eries together into one mechanism, and it has not been proven.

The rate at which we grow is far from constant; we put on a growth spurt at puberty that is unique to Man, having apparently arrived at the *Homo erectus* stage of our evolution.[23] After this spurt, our skeletons stop growing although our bodies may increase in girth with added muscle or fat. Both phenomena, the spurt of growth and its cessation, are apparently driven mainly by the sex hormones that trigger the other obvious effects of puberty such as growth of body hair and mammary glands.[24] For growth, the most important hormone in this respect is oestrogen.[25] This is often thought of as a 'female' hormone because of its role in the menstrual cycle, but men also make it thanks to an enzyme that converts testosterone to oestrogen. Oestrogen stimulates the production of Growth Hormone and drives growth faster and also affects bone cell behaviour directly.[26] Growth can be so fast that it slightly outstrips the speed at which minerals can be laid down to make fully mature bone and the bones of teenagers can become very fragile. Nearly half of all children break a bone during adolescence,[27] and half of those breaks are in the rapidly growing long bones of the arm (although fragility of bone is probably not the only reason for this life-time peak in fracture risk: there is also the time gap between a young lad's acquiring strength and his acquiring the common sense to go with it).

From the point of view of the growth plate, there is a price to oestrogen-mediated fast growth. High concentrations of oestrogen encourage cells to stop proliferating to make cartilage. This effect is so strong that it seems to override the delicate balance of the self-organizing feedback loops that maintain the organization of the growth plate. This means that, by the very end of puberty, when oestrogen levels are very high in both sexes, proliferation fails to keep up with maturation. Gradually the entire growth plate is converted to mature cartilage. Once this has taken place, the growth plate is 'closed' and its ability to grow is exhausted. Some people carry mutations that mean that they cannot make oestrogen: such people never close their growth plates and they continue to grow as adults. This adult growth can be stopped by giving oestrogen injections, which close the growth plates down. Indeed, even 'normal' girls who are growing unusually tall have been treated with oestrogen to limit their final height. Similarly, short adolescent boys have been treated with drugs that block oestrogen action to hold their growth plates open for longer so that they grow taller. Arguments about the ethics of such an intervention, and the question whether anyone has a right to say what height another human being should reach, belong in a book other than this one. For our purposes, there are two clear messages: oestrogen closes growth plates, and altering the timing of growth plate fusion can alter final height. This timing is therefore the other major factor that, along with the speed of growth, determines how tall someone grows.

To sum up so far, we have the following tentative explanation for size control in the skeleton: (i) bones grow by means of self-organizing patterns of proliferating and maturing cells, which signal to one another; (ii) circulating Growth Hormone encourages growth, but the more a bone has grown, the less sensitive it becomes; (iii) sex hormones drive rapid growth at puberty, but when they rise very high they disturb the self-organization of the growth plates so that the plates close and growth ceases. Growth Hormone is made directly in the pituitary gland, while sex hormones are made in the gonads in response to commands coming mainly from the pituitary. In this sense, while the skeleton is the master tissue of growth, the pituitary gland is in overall charge.

What of the 'slave' tissues—those that match their own size to the general body size achieved by their skeletal 'master'? It is striking that in normal animals and in mutants with all sorts of abnormalities in skeleton size and proportion, the slave tissues grow appropriately for the body they have to serve. This is even true when an unusual body shape means that the slave tissues have to achieve unusual dimensions, as in the skin of a limb that is the normal breadth but very short, or the skin over the belly of someone seriously obese. Given this robustness, it seems unlikely that the 'slave' tissues simply follow biochemical signals' reporting of how much master tissue is present, because there is no obvious way that such signals could report the shape of what is needed, especially in abnormal situations. There is, though, one type of signal that would report both relative size and shape very effectively at any time of life and for any body shape, however bizarre. That 'signal' is mechanical force.

If a tissue—the skin of the leg, for example—were not keeping up with the growth of the underlying tissues, it would become stretched and would experience tension. Tension is a property that exists throughout an object, and a 10 per cent stretch of a tissue would be detectable as a 10 per cent stretch at all points within it, however large or small the tissue already is. A mechanism that used excessive tension as a reporter of inadequate growth would have the merit of size-invariance: it would work just as well for a large tissue as for a small one. It would also have the merit of having to carry no expectation or knowledge of shape. As long as cells under excess tension proliferated by dividing to make daughters that added extra tissue area in the direction of that tension, tissue would always be added in the correct direction without anything having to 'know' in advance which way to grow. The system would therefore be completely robust over a range of body shapes, making accommodation of unusual growth patterns easy and, incidentally, making evolutionary change of body shape easier too.

There is strong evidence that mechanical tension is able to drive proliferation. If a gentle force is applied to the ear of a living rat for a few days, the cell proliferation in the ear increases and the tissue grows.[28] The distended ears of people who habitually wear heavy earrings suggest that this phenomenon occurs in humans too. The encouragement of cell proliferation by tension can be demonstrated in simple cell culture systems that put some cells under more tensile stress than others.[29, 30] The culture systems consist of shaped 'islands' of cell-friendly surface, surrounded by surfaces to which cells cannot attach. The cells that are applied to these islands are the types of cells that form sheets and tubes, and adhere to each other with cell–cell junctions. Inside the cells, the junctions are connected mechanically to one another by the protein microfilaments that run through the inside of the cell, and always exert gentle tension. The shapes of the islands, which can be as simple as squares or stars, feature straight edges and sharp corners. Cells lying along straight edges suffer no

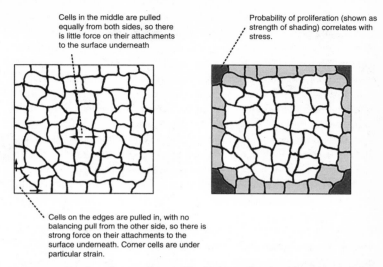

Cells in the middle are pulled equally from both sides, so there is little force on their attachments to the surface underneath

Probability of proliferation (shown as strength of shading) correlates with stress.

Cells on the edges are pulled in, with no balancing pull from the other side, so there is strong force on their attachments to the surface underneath. Corner cells are under particular strain.

FIGURE 77 Growing cells on small square 'islands' places cells at the edges, and particularly at the corners, under strong mechanical stress. These cells proliferate more than the unstressed ones in the middle of the island (this figure is based on an original micrograph by Celeste Nelson and colleagues[29]).

strong mechanical effects from the edge, and experience just the gentle tension that all cells make and feel. Cells at sharp corners, though, have to bend awkwardly and still resist the gentle pull from their neighbours. As a consequence, they are under significantly greater tension. It is these cells, at the corners, that show most proliferation (Figure 77). If the surface on which the cells are growing is itself stretched, something that would mimic a tissue being stretched by not keeping up with the growth of the body around it, proliferation takes place throughout.

As tension and stretch might be used to tell a tissue that it is not growing quickly enough, compression and crowding might be used to tell cells that they have already proliferated quite enough and should cease. Many years ago, it was observed that when normal cells are placed on the bottom of a culture dish, they proliferate until they have covered the bottom of the dish and then stop.[31] If some cells are then removed to create space, for example by wiping a sterile pencil eraser across the dish, the cells surrounding the gap proliferate again until the hole is filled, and then they stop again. This phenomenon, termed 'contact inhibition', provided an early hint that the proliferation of cells may be controlled by their own sense of crowding. For many decades, the mechanisms proved elusive, but very recent data from fruitflies have cast some light on the problem. Key to the mechanism as we currently understand it are two large cell surface proteins that are structurally similar to cell adhesion molecules that stick cells to one another; indeed there is very strong evidence that the molecules do stick to one another. The more crowded cells are, the more of their surfaces will be pressed together and the more interaction there will be between these surface proteins. Their interaction triggers a complicated signalling pathway inside the cell, which ends up inhibiting proliferation.[32, 33]

Another way in which a slave tissue might sense the appropriateness of its size is to sense its own biochemical effect on the body, either directly or indirectly in conversation with other tissue types. This type of regulation has already featured in Chapter 9, where the growth of blood vessels was controlled in part by how good a job the existing blood supply was doing of providing tissues with oxygen. The possibility that whole organs may use a method like this is suggested by transplant experiments in which a non-vital organ is removed from an animal and then one or more foetal organ rudiments of that type are grafted back in. Generally, these experiments are done to learn more about transplantation, and the data on size control has been an added benefit.[34] If one foetal spleen rudiment is grafted into a host whose own spleen has been removed, it grows to the size of a normal adult spleen.[35] If multiple spleen rudiments are grafted into the same host, then each stops growing when the sum of their individual volumes has reached the size of an adult spleen. This suggests that either the spleens themselves detected when there was enough spleen-type tissue somewhere in the body, or the body detected the fact and signalled it to the little spleens. The phenomenon is not universal, though. When approximately the same experiment was been done with the thymus, each rudiment grew to the size of a normal thymus, leaving the animal very over-endowed in that department.[36] It is clear, therefore, that different organs can regulate their sizes according to different rules. This makes the whole field a lot more complicated.

Much remains to be discovered about the control of size and proportion, but this chapter has outlined some of what is known. The skeleton acts as a master growth tissue to set the size of the body, and most other tissues follow it. The skeleton's growth is regulated by the pituitary's secreting growth and sex hormones, and these are interpreted by special self-organizing growth plates within growing bones. Other tissues respond to mechanical stresses placed on them by skeletal and other growth, and expand accordingly. Some internal organs may also sense their size relative to the body by sensing their biochemical effects, though not all seem to. None of these mechanisms requires any cell to have any picture of the tissue in which it is located. Neither does a cell require a blueprint or a detailed list of instructions, beyond simple rules such as 'proliferate faster if you receive such and such a signal'. The size of the body, the proportions of its parts, and its symmetries all emerge from the actions of these simple, blind, local rules. In this respect, the body of a teenage boy, containing far more cells than there are stars in our galaxy, uses the same basic principles to control its development as the body of the tiny embryo that he once was.

17

MAKING FRIENDS AND FACING ENEMIES

Support bacteria—they are the only culture some people have.
car bumper sticker

The mind is not the only learning machine that develops in the years following birth. Development of ways to face, and to use, an environment full of microorganisms is critical to the health of a growing human.

We are never alone. For the first nine months of our lives, we live inside our mother. After birth, we share our bodies with about a hundred trillion micro-organisms, even when we consider ourselves to be clean. This number is so large that micro-organisms outnumber the human cells of our bodies by about ten to one.[1] When we die, these organisms live on to feed on our remains and on each other until there is nothing left. Some of these microbes are mere fellow-travellers, doing us neither good nor harm, but many are important to the way that our bodies work, and we need them because they can perform biochemical tricks that human cells have never evolved.

The healthy gut contains between one and ten billion bacteria per gram of tissue. These tiny organisms perform several important tasks, some of which will be described later in this chapter and one of which will be introduced now: the bacteria secrete a variety of enzymes that can digest components of food that our own enzymes cannot attack.[2] The enzymes break down large, awkward molecules into small pieces that can be absorbed by the gut lining and by bacteria, both of which regard them as food. The bacteria consume the molecules they absorb there and then, and use the energy and raw materials to multiply and to make more enzymes. The gut lining passes the food to underlying blood vessels, from which it travels to the liver for processing and then on to the rest of the body. Some of the food molecules attacked by bacterial enzymes would be toxins or carcinogens if left intact, so a second important function of gut bacteria is rendering food safer.[3] They are also responsible for making some 'foods', particularly alcohol, more dangerous (in the case of alcohol, by metabolizing it to acetaldehyde, a toxic and probably carcinogenic derivative).[4] Some gut bacteria produce significant amounts of vitamin K, which is important in blood clotting and bone growth, and which human cells cannot produce for themselves.[5] Some also make folic acid to add to that already in food; adequate folic acid is important in human cell proliferation, and the mother's having enough folic acid is particularly important in neural tube closure during early development (Chapter 5). Given their participation in digesting food, gut bacteria have plenty of access to energy and raw materials and multiply quickly. Most are, though, removed as the remains of food travel onwards: around three-fifths of the mass of a normal human stool consists of bacterial cells, many dead and some still living.

The vagina is another environment rich in useful bacteria. Here, as in other areas potentially vulnerable to pathogens, they are needed to make the environment inhospitable for other micro-organisms that might otherwise take advantage of the warm, damp area to set up an infection. Vaginal *Lactobacilli* bacteria feed on components in mucus and secrete lactic acid that is too strong for most other microbes to tolerate, and also inhibit other microbes in other ways.[6] The vagina, in which one set of microbes helps defend against others, illustrates a general problem that also applies to the gut and to many other sites: the body has to find a way of supporting helpful bacteria while not laying itself open to invasion and attack by dangerous, pathogenic ones. Recent research suggests that humans and their helpful bacteria have, over their long association, evolved ways of communicating with each other so that the two very different types of organism can operate as a single, integrated system.

The environment in which the foetus grows, deep in the womb and surrounded by several layers of membrane, is sterile. A new human therefore forms without bacterial partners, and has to acquire them at, or shortly after, birth. Fortunately, the position of the birth canal, summarized in St Bernard of

Clairvaux' observation, *inter faeces et urinam nascimur*,* is ideally suited to making sure that a baby will come into contact with bacteria from the normal human vagina, gut, urinary tract, and skin at the very moment of its somewhat messy entrance to the world. Even infants delivered by caesarean section will meet these bacteria once their mothers start to handle them, although they can take significantly longer to develop their normal bacterial flora.

When symbiotic bacteria enter the mouth of a new baby, they are swallowed along with saliva or milk and they pass through the stomach to the intestines. Here, they begin their signalling dialogue with the human cells. The process has been studied more in mice than in humans, but parts of the cellular story have now been checked at least in human cell cultures, and epidemiological studies in humans suggest that humans and mice are likely to be broadly similar in the way they manage these things.

Each part of the gut creates a nurturing environment for precisely the type of bacteria that it needs. This can be illustrated by the interactions between the gut cells and one of the most important symbiotic partners of mice, the species *Bacteroides thetaiotamicron*. Once in the digestive tract, this bacterium secretes a molecule that is detected by the lining of the small intestine. The small molecule is very different from the proteins described in previous chapters that human cells typically use to signal to one another in development, and is related much more to the ordinary biochemistry of bacterial metabolism. Nevertheless, this chapter will follow the principles of biosemiotics[7] and insist that any molecule that carries information about some state of affairs, and that affects the behaviour of a cell that receives it, should be regarded as a 'signal', whatever its molecular nature and whatever the primary reason for its being made.

In response to the signal, intestine cells make a subtle change to their own metabolism. Animal cells in general, including those of humans, tend to decorate the proteins they secrete or place on their surfaces with chains of sugars. Within the chains, the sugars are linked to each other with strong chemical bonds and, unlike single molecules of sugar floating about freely in a glucose drink, for example, they cannot be used directly as food because the sugar-protein complex is far too large to pass through the uptake channels on the surface of a bacterial cell. The only way that a bacterium can get food from these sugar chains is by having an enzyme that can cut the bonds holding the chain

---

* This quotation, meaning 'we are born between faeces and urine', is often mis-attributed to Augustine of Hippo, probably because he is the saint most associated with dry one-liners.

together, and different enzymes are needed to liberate different types of sugar from the end of a chain. *Bacteroides thetaiotamicron* makes an enzyme that can liberate the sugar fucose from the ends of one of these chains. Before they receive the signal from *Bacteroides thetaiotamicron*, intestine cells do not place fucose residues at the ends of many of their chains but, once they receive the signal, they switch to add the fucose.[8] Effectively, the bacterium is saying 'feed me' and the cell is obliging, and doing so in a way that will not feed other random organisms that may be present but that will not have the right enzyme to liberate fucose. Other parts of the gut express other types of linked sugars, probably in response to other signals from other symbiotic bacteria. In this way, each part of the gut might create a nurturing environment for the bacterial species it needs.

The advantage to the gut of having *Bacteroides thetaiotamicron* is that it can process hard-to-digest components of food and liberate nutrients to be taken up by intestinal cells. Uptake of food, or any other molecule, is always most efficient if the molecules already taken up are cleared away to distant parts of the body, otherwise some tend to leak back out again and be wasted. The efficiency of this clearing-away process is improved by another response to signals from symbiotic bacteria.[9] These signals are detected by an intestinal cell type that, in response, causes blood vessels to produce a dense bed of branched capillaries, by the mechanisms described in Chapter 9. This network helps to collect absorbed nutrients and send them on to the liver and thence to the rest of the body, preventing them piling up in the intestinal wall and leaking back out again. Yet other signals[10] from *Bacteroides thetaiotamicron* induce tissue cells to secrete antibacterial molecules that are relatively harmless to this particular bacterium but are toxic to some unwanted and dangerous competitors such as *Listeria*. The symbiotic bacterium and the gut lining therefore look after each other, the body providing food and protection for the bacterium, the bacterium providing nutrients for the body, and their combined action discouraging colonization by less friendly species. This discouragement proceeds both by the system making antimicrobial molecules and by the *Bacteroides thetaiotamicron* simply taking up the space that would otherwise be available for unwanted species (Figure 78).

Not all bacteria are welcome in the gut, or anywhere else in the body. Countless thousands of bacterial species would invade and consume our warm, nutrient-rich bodies if we allowed them to, and even normally symbiotic species can cause disease if they manage to enter the tissues themselves (for example,

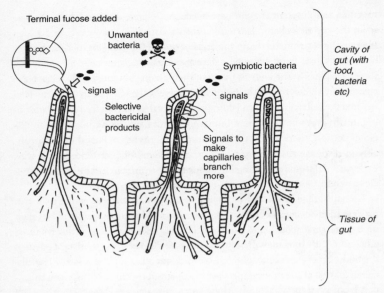

FIGURE 78 Summary diagram of important signalling events between symbiotic bacteria and gut tissue, and their consequences. Signals from the symbiotic bacteria cause the gut cells to add fucose residues to their sugar chains, to make internal signals to drive capillary growth, and to secrete compounds that kill rival bacteria. In this diagram, the size of the bacteria, in relation to the tissues, has been greatly exaggerated, as has the space between the bacteria and the tissue.

when a bullet wound, an ulcer, or a cancer perforates the intestine wall). 'Friendly bacteria' are friendly only if they can be kept in a safe place. Fortunately, we are equipped with elaborate defensive systems that are effective in killing most invaders. The few bacterial species that have evolved to slip past these defences, at least occasionally, have been responsible for some of the notorious diseases that have shaped, and in some parts of the world continue to shape, human history. They include tuberculosis, bubonic plague, leprosy, syphilis, diphtheria, cholera, and typhoid. To understand how we can both nurture wanted bacteria in their correct place and at the same time have defensive systems that kill bacteria of certain types, or in a place in which they are not wanted, we need to examine how our defensive systems work.

There are three basic layers of defence. The first layer is essentially passive, and it has both physical and chemical components. The physical component consists of barriers to infection, such as tough layers of dead cells on the outside of the skin, frequently-replaced, viscous mucus inside the nose, mouth, trachea, gut, urethra, and vagina, and even films of symbiotic bacteria that are impenetrable to pathogens. The chemical component consists of a variety of proteins that bind to and damage the cell walls of bacteria. These wall structures are unique to bacteria—animals cannot make them—so they form a useful target that can be attacked chemically without risking the health of tissue cells. Animals therefore make a variety of enzymes and pore-forming proteins that make lethal holes in bacterial walls. Some, including lysozyme and defensins, protect external surfaces such as the eyes. Others, such as the pore-forming proteins of the complement system, are found throughout the blood and the fluids that bathe internal tissues, and can be activated directly by pathogen surfaces as well as by other means to be described later in the chapter. Together, the passive systems form an ancient first line of defence, found in broadly similar forms right across the animal kingdom.

The second layer of defence is active and uses various types of migratory cells called phagocytes, literally 'cells that eat'. There are several different types of phagocyte, but all have their origins in the bone marrow, a tissue to be considered in more detail in Chapter 18. They spread through the body in blood but can choose to leave blood vessels by pushing between the cells that line them. Once across the blood vessel wall, the cells enter the tissue spaces where they can settle or actively patrol, crawling through fine spaces. When they are crawling, phagocytes have much in common with the migratory cells of the embryo that were the focus of Chapter 8. They make a leading edge, the extension of which is controlled by the activity of signalling pathways triggered by receptors on the cell surface. Phagocytes have receptors that are triggered by two broad classes of signal. One type of signal, of which there are numerous different examples, consists of bacterial cell walls and waste products that living bacteria cannot help releasing. As far as the phagocytes are concerned, these are also signals; signals that mean bacteria are nearby. The other type of signal, of which there are again many examples, consists of molecules released by stressed and dying human cells.

The point of the receptors that recognize bacterial molecules is the easiest to understand. In the same way that migrating embryonic cells navigate towards the source of embryonic signals (Chapter 8), phagocytes migrate towards bacteria. Once there, the phagocytes secrete more signalling molecules, release a cocktail of highly toxic chemicals, and endeavour to engulf and destroy any bacteria they meet. The signalling molecules increase local blood flow and leakage of fluid from blood vessels, bringing more phagocytes as reinforcements. The toxic chemicals kill bacteria even without their having to be engulfed. They are so aggressive that they often inflict considerable damage to the ordinary human tissue as well. The result of all of this is a local build-up of redness and heat, from the increased blood flow, swelling from the fluid and accumulating cells, and pain from nerve endings suffering under the onslaught of a toxic cocktail: *rubor, calor, tumor et dolor*, the classic signs of inflammation described by Celsus nearly two thousand years ago. In the middle of the inflammation may also be an area of whitish pus—essentially phagocytes, dead bacteria, and dead human tissue. The skin pimples that many people suffer around puberty, which are caused by bacterial infection of sebaceous glands in the skin that have become blocked by excessive production of sebum under hormonal stimulation, show these signs very clearly. In the case of acne, the inflammation is a nuisance but, in the context of a dangerous infection, it can be a life-saver, bringing the full force of the body's defences to bear on the cause.

The point of receptors that recognize signals from stressed human cells, and treat even the detritus of dead human cells as signals, is flexibility. Not all dangerous micro-organisms are bacteria: some are unicellular, animal-type cells, with no alien cell walls to provide an easy target, and fewer peculiar waste products to announce their presence. Some parasites, such as worms and flukes, are complete small animals in themselves, with biochemistries very similar to our own. Viruses, the smallest of our parasites, are essentially just tiny parasitic genomes wrapped up in coats made by infected human cells, and therefore appear to be very human in terms of their basic chemistry. Having a defensive system that relied completely on recognizing particular chemical signatures, such as bacterial wastes, would leave us completely open to attack from anything that did not make those signature chemicals. Even if we evolved more and more receptors to try to keep up with more and more types of pathogen, the chances are that the evolution of the pathogens, which tend to reproduce very quickly and inaccurately to produce immense populations with considerable variation, would almost certainly outrun our own. We are forced into an evolutionary arms race but, handicapped by our slower breeding, we therefore need some way of directing an active response to anything that is causing injury, even when it is not detectable directly. This is where receptors that detect

MAKING FRIENDS AND FACING ENEMIES

stressed and dead human tissue come in. Anywhere cells are being killed,[†] a defensive response will be mounted. That is why even sterile burns are followed by painful inflammation. In the context of sudden tissue damage, the apparently excessive violence of excited phagocytes, whose toxins kill not just bacteria but even neighbouring human tissue as well, makes sense: even a micro-organism or virus that hides inside those cells and so could never be detected or tackled directly by the phagocyte, will be killed in the general melee. The loss of normal tissue can be serious, but at least the disease is stopped in its tracks.[‡]

Damage to human tissue is therefore an important general controller of the body's active defensive response.[11] This already provides a hint about how we tolerate symbiotic partners while at the same time fighting off other bacteria, but a fuller explanation requires an appreciation of the third layer of defence. This is something possessed only by vertebrate animals: the ability to learn from experience. It still uses the ancient weapons of complement—the soluble chemical-based defence—and phagocytes to kill invaders, and it is still controlled, ultimately, by signals from damaged or infected tissue. What it adds is a few more cells and, critically, a set of highly specific proteins that can direct phagocytes and complement very accurately and rapidly against an invader, provided that the system has enough time. Because of its ability to learn and adapt, this new layer is called the adaptive immune system.

The adaptive immune system, consisting as it does of migratory cells in relatively brief contact with one another, has a physical structure very different from that other learning machine, the brain. Nevertheless, the underlying logic is remarkably similar. In the brain (Chapter 15), the basic principle of learning was to begin with a large number of connections, some of which would turn out, in the light of experience, to be appropriate, and some of which would not. The signalling traffic generated by experience itself acted to eliminate inappropriate connections and to strengthen useful ones. In the adaptive immune system, traffic passing through receptors plays a part analogous to connections between neurons. Again, the principle is to produce a very large

[†] The elective cell death described in Chapter 14, which takes place as a normal developmental process, does not involve the release of stress signals, and the remains of the cell are cleared away in such a way that no alarming traces remain. For this reason, developmental cell suicide does not excite defensive systems.

[‡] Usually: some diseases actually use the self-destruction of the body to help them to spread through tissues.

number of different possibilities—receptors, in this case—some of which will be appropriate and some not, and to use the signalling traffic through them to decide which will have their production strengthened and which will be eliminated.

The overall strategy for learning may be similar, but the details are very different because the adaptive immune system is fluid, with no fixed physical architecture, and cannot rely on discrete point-to-point connections as the nervous system can. Receptors critical to learning are borne on a type of cell unique to vertebrates, the T cell ('T' for thymus, a gland where these cells spend much of their early lives). T cells come in various types; some types control the activities of other cells, while others can act as executioners that inject lethal enzymes into tissue cells that have been identified as infected. All carry the critical T-cell receptors, or TCRs. For the learning system to work, it needs to start with millions of T cells. Each cell carries just one of a vast range of TCRs, and each TCR is unique in its preference for what it will recognize.

Generating this diversity of TCRs poses an interesting problem. The preference of a TCR for binding to specific other molecules depends on the precise sequence of amino acids in the proteins that contribute to the TCR.[12] As with all proteins, the sequence of amino acids is determined by the sequence of bases in the gene. In principle, an animal could gain the ability to produce a few varieties of receptor by having several versions of the gene, each subtly different from one another in terms of base sequence. This approach is indeed used in many families of signalling receptors used in development, but for the adaptive immune system, it will not do. We have only around twenty-five thousand genes, yet the adaptive immune system requires millions of different TCRs to be produced. Adding millions of new genes to the genome is out of the question for many reasons: they would be far too expensive, in terms of raw materials, to keep copying as cells divide, they would make the genome unstable because of a process called recombination, which takes place between similar stretches of DNA and, in any case, this much DNA would not fit in a cell in the first place.

T cells solve this impasse by having just one basic gene for each of the two types of protein chain found that come together to make a TCR, and then displaying the most outrageous disregard for the normal 'rules' of biology: each cell deliberately mutates and re-arranges part of the gene for the TCR. They have a special set of enzymes that rearranges *only* this region of *only* this gene, in a process that involves several complicated steps, the upshot of which is that each T cell makes a TCR from genes that have an essentially random base sequence in just the part that specifies what the TCR will bind. Each individual cell can make only one version of them.

The downside of this randomness is that many of these versions of the TCR will be incapable of binding to anything, even weakly, and will therefore be useless. Some will have potential to recognize pieces of dangerous micro-organisms on cells and might be valuable in defence. Others will recognize the normal tissues of the body, an activity which is at best useless and at worst positively dangerous. The first aspect of learning in the adaptive immune system is therefore the weeding out of useless or dangerous TCRs, which means the weeding out of the T cells that bear them.[13] Young T cells that have just finished the random gene rearrangement and have started to make their TCR live in the thymus, surrounded by cells that have many fragments of body protein on their surfaces. The TCRs have plenty of opportunity to recognize the fragments if they happen to fit the TCR. The life of each T cell depends, at this stage, on its TCRs being stimulated, but only weakly. A TCR that reports absolutely no binding to anything is probably useless, and the cell that carries it kills itself. A TCR that reports weak, sporadic binding can clearly detect fragments of molecules, but not strongly. This is actually quite hopeful, because detecting a fragment of a normal body protein weakly means that the TCR at least works, and it *may* detect some unknown bacterial or viral protein fragment much better. These cells therefore live, mature, and enter the rest of the body. A TCR that reports strong binding while still in the thymus is almost certainly recognizing fragments of normal human material and either kills itself or enters a state that suppresses activity against those fragments in order to prevent the immune system acting against its own body.

At the end of this process, the body will have populated itself with millions of T cells, all carrying different versions of the TCR, none of which react strongly to body components, but all of which have shown a very weak response to at least something. While out in the body, T cells have frequent contact with types of phagocyte that present them with little fragments of what they have engulfed, again on fragment-displaying proteins at the cell surface. If the phagocyte has come from a site of infection, there will be pieces of the microorganism amongst these fragments. For a given micro-organism, the TCRs of most T cells will not recognize anything but, occasionally, the phagocyte will happen to present its load to a T cell that does recognize it. This is what the T cell has been waiting for, and it becomes fully activated. It proliferates quickly to create a small army of daughter cells with the same TCR, and it secretes signalling

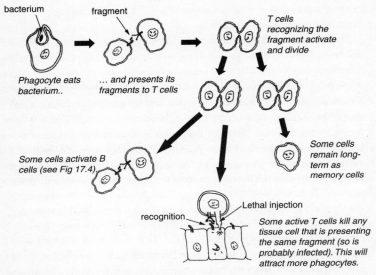

FIGURE 79 Summary of some of the ways that T cells, with TCRs that are activated by a fragment being presented by a phagocyte, organize and take part in a defensive response. Various events in this diagram, such as cell death and complement, can also recruit the phagocytes involved in the second level of the immune response.

molecules that recruit further T cells to the area. Some of these will recognize other fragments of the microbe and they too will become activated. A subset of these T cells will go on to kill tissue cells that display the same protein fragment, and are presumably therefore infected (Figure 79).

The learning aspect of all of this depends on both the cell multiplication involved, which means that there are now many cells all carrying that exact type of TCR, and the ability of some of these cells to persist, after most of their sisters die off. The surviving cells become memory cells. Their presence means that, if the same pathogen invades again, there will be far more cells to recognize it and trigger a response. Moreover, the memory cells are more sensitive to that pathogen than any cells were in the first invasion, and they alter the way they communicate with the rest of the immune system. The result of this is that a repeat challenge from the same pathogen results in a very rapid and efficient defensive response. Under these circumstances, the invader stands very little chance of winning. For this reason, we suffer most diseases only once in our lives and are then immune to them, even when being coughed over or touched by sufferers. The apparent exceptions, such as the common cold, are actually caused by a variety of different organisms (viruses, in this case) so each time is like the first. A few micro-organisms, such as the causative agent of malaria, are an exception to this because they have evolved a range of mechanisms to evade the immune system: this is an example of the evolutionary arms race alluded to earlier in the chapter.

T cells are not the only cells of the body to use random rearrangement of specific genes to produce a vast variety of receptors. A similar set of cells, B cells (which develop in the bone marrow§), use exactly the same trick of gene rearrangement to make B cell receptors, or BCRs. These are very similar to TCRs. Again, each individual B cell has just one unique variety of BCR. B cells patrol the body and, if they find any molecule that their BCR recognizes, they take it in and chop it up with enzymes. They then present its fragments on their surfaces, on a fragment-displaying protein, on the off-chance that they might meet a T cell that will recognize the complex. If they do, the T cell signals back to the B cell, causing it to multiply to make more B cells with the same BCR. Some daughters of the B cell become memory cells, in readiness for future battles against the same microorganism, while others start to secrete their BCR into the fluid around them (Figure 8o). Secreted BCR is called antibody, and it can spread rapidly through tissues and bind to the molecule it recognizes, whether that molecule is floating about or still on the surface of the microbe or infected cell. Antibody recruits that ancient chemical defence, complement, and also phagocytes: it is thus a death warrant for any cell it binds.

The learning response of the adaptive immune system is the basis of the medical technique of vaccination. Here, particular proteins from a dangerous bacterium or virus, or a harmless or weakened strain, are injected to create a first battle. When it is over, the body has memory T and B cells that can mount a rapid and effective response if they ever meet the real disease. Effective vaccination usually depends on injection of a preparation that is irritating enough to cause tissue damage, to recruit cells, and to help the target molecule persist long enough to be recognized: pure, clean protein does not work well.

---

§ The 'B' of B-cell actually stands for Bursa of Fabricius, an organ present in birds but not in mammals, in which B cells are made; it is a lucky coincidence that the letter is also appropriate for the bone marrow, where mammals make their B cells.

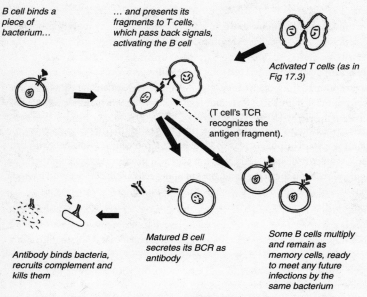

B cell binds a piece of bacterium...

... and presents its fragments to T cells, which pass back signals, activating the B cell

Activated T cells (as in Fig 17.3)

(T cell's TCR recognizes the antigen fragment).

Antibody binds bacteria, recruits complement and kills them

Matured B cell secretes its BCR as antibody

Some B cells multiply and remain as memory cells, ready to meet any future infections by the same bacterium

FIGURE 80 Diagram of B cell activation. If B cells bind a piece of bacterium that their BCR recognizes, they present it to T cells. If any B cells find a T cell whose TCR recognizes the fragment, the cells communicate and the B cells become activated. Some multiply and remain as memory cells that can be activated quickly in a future infection. The rest mature and secrete their BCR as antibody. The antibody, which retains the original BCR's ability to bind to that specific type of bacterium, can recruit complement and phagocytes when it does so, and therefore cause the rapid death of that bacterium.

This highlights the way in which the initial activation of the adaptive immune response rests ultimately on the ancient production of alarm signals from stressed and dying tissue as well as from bacterial products.

The adaptive immune system has no need for a priori, genetically determined knowledge of the chemical details of the adversaries it will face. Like the brain, it is altered by experience, and T and B cells that have had their receptors activated in a battle will remain as a rapid reaction force, ready to strike again at that particular foe without even waiting for significant tissue damage to taken place. In this respect, what does not kill us really does make us stronger.

An outline understanding of the adaptive immune system brings us back to the question of how we tolerate symbiotic bacteria. Very recently, it has been discovered[14, 15] that human gut cells that detect signals from symbiotic bacteria produce signals to the defensive systems that effectively say 'nothing alarming is happening here'. The signals act on the phagocytes that specialize in presenting surface-bound fragments to T cells. These phagocytes** can be in two states: in one state, they present the fragments along with signals that encourage T cells to commence an aggressive attack, while in the other state they accompany their presentation with signals that encourage the T cells to calm down and be tolerant. Unstressed gut cells in contact only with symbiotic bacteria secrete two proteins that instruct the phagocytes in their vicinity to enter the calming state. The phagocytes will still present molecules associated with bacteria, with part-digested food, etc., but they will promote tolerance rather than aggression. If, on the other hand, the gut lining cells are in contact with bacteria that damage them, they do not make these calming signals and make alarm signals instead. Under these circumstances, phagocytes present fragments to T cells along with strong activating signals, and a defence is quickly mounted. Again, the main controlling element of the defensive response is whether or not a tissue (including the immune cells therein) is suffering stress. What the dialogue between bacteria and gut lining adds to this is a positive message that nothing bad is happening.

The influence of symbiotic gut bacteria on the immune response may go far beyond ensuring their own survival. Mice that are raised in sterile conditions and have no gut bacteria have peculiar and defective immune systems,[16] vulnerable to attack from various micro-organisms including those that have nothing to do with the gut. It now seems that fragments of the surface of some gut bacteria, including *Bacteroides fragilis* which is one of the earliest colonizers after birth, profoundly affect the maturation of different populations of T cells in the whole body, altering the balance between those that act to drive immune responses and those that act to encourage tolerance. If the normal development of our immune systems depends on contact with harmless bacteria (probably hundreds, and not just *Bacteroides fragilis*), this may explain why increasingly good hygiene, which is generally regarded as a good thing, has been accompanied by an increase in the incidence of diseases such as asthma, which reflect an immune system out of balance and inclined to be excited by harmless substances such as dust, animal hair, and pollen.[17] Another hygiene-related change

---

** This is a distinct type of phagocytic cell called a dendritic cell.

has been the near-abolition of infections by intestinal worms. Humans and their worms have evolved together for a long time and it turns out that worm infections alter the balance of the immune system, essentially calming it down to the benefit of the worm and also to the host. Worm-free animals, and probably people, show much more immune hyperactivity than animals with modest numbers of gut worms (excessive numbers of worms create problems of their own, of course).

In humans, the two main learning systems of the body, the brain and the adaptive immune system, have begun to work together to add a fourth layer of defence to the three possessed by other vertebrates. Even in other mammals, instinctive and learned grooming behaviours, and avoidance of food that smells rotten, allow the brain to assist defensive systems by limiting the individual's contact with dangerous microbes in the first place. Human brains, with their ability to investigate the world systematically and pass knowledge on to future generations, add a whole new dimension to behavioural defence. Two of the most dangerous things we have to do, in terms of laying ourselves open to infection, are eating and drinking, because there is always a risk that food and water harbour hidden pathogens such as *Salmonella* or *Cholera*. Our ancient cultural inventions of cooking using fire, and of drinking boiled or fermented liquids instead of cold water, cut down the risk of infection dramatically. It is probably no coincidence that all civilizations that developed towns with large, concentrated populations of people living together, had already invented tea, ale, or an equivalent. More recently, discovery of bacteria and the routes by which they infect people, and consequent construction of plumbing to bring clean water to people and remove sewage from them safely, allowed the construction of vast cities with stable populations of millions. Development of vaccines, antibiotics, and anti-viral compounds has more recently still added significant safety at the level of individuals as well as populations. There will always be new threats: pathogens evolve quickly and our global travel lets them spread like wildfire. Our populations are already so high that we are by now utterly dependent on having this fourth, cultural/scientific level of defence. If, as a species, we ever turn away from painstaking scientific endeavour, we will pay a terrible price.

Development of the mind, considered in the last chapter, and development of the immune system described here are each critical developmental events that take place after birth. They have to, because both involve interaction of the new human with an unpredictable environment. Their timing emphasizes the fact that birth, while dramatic, is by no means the end of human development. Indeed, both the mind and the immune system continue to develop for life, and are always responsive to new experiences. As its life-long development proceeds, the body also has to maintain its structures against injury and wear-and-tear: that maintenance, and its relationship to foetal life, will form the topic of the next chapter.

18

MAINTENANCE MODE

' You're the worst kind; you're high maintenance
but you think you're low maintenance.'
Nora Ephron

One of the games that biology teachers like to play with their students is to ask them to define life; to come up with a simple criterion that could be used to determine whether some entity is living or not. The game can be played at all levels, from elementary school to doctoral student and beyond, and at each level it stimulates the same basic debates (although doctoral students tend to use—sorry, 'utilize'—unnecessarily long words to try to make their arguments sound more impressive).

Even elementary school children can quickly dispose of spurious criteria such as whether something moves under its own power, pointing to living things that do not move, such as corals, and non-living things that do move, such as raindrops. Similarly, they can dispose of criteria such as reaction to stimuli such as being poked, by considering living non-reactors such as toadstools, and non-living reactors such as mouse-traps. The ability to reproduce is often given as a defining characteristic of life, even by the authors of college textbooks who really should know better, but this idea is again easy to demolish: to take it seriously would be to consign red blood cells, mules, worker ants, and post-menopausal women to the world of the non-living. More advanced students may offer self-organization as a criterion, but again there are non-living entities, such as crystals, convection cells, and the wave patterns of some chemical reactions (e.g. Belousov-Zhabotinsky*), that show at least some self-organizing activity.

There is one feature of the living world that does appear to be universal, however, and that is the ability of living things to use externally obtained energy to maintain and regenerate themselves from within. This criterion, stated in that form by Pier Luigi Luisi,[1] builds on previous definitions suggested by Aleksandr Oparin and Jacques Monod (each of these scientists has made outstanding contributions to understanding how life might first have arisen). The requirement for repair processes comes partly from the incidental external damage to which all organisms are vulnerable, but also from the inherent fragility of the materials from which we are made. This fragility means that, even in the absence of our being hit, bitten, scraped, or blown about, our constituent molecules last a relatively short time in the chemical melee of a living cell and have to be replaced frequently. The same fragility may be present in man-made objects, with individual components having to be replaced many times during their lives, but the difference is that the man-made objects are not able to make and to fit new components for themselves. If my car requires brake shoes, I have to source them from an external supplier and fit them manually; supplying the car with raw friction material, or even with complete unfitted shoes will not, alas, result in the car replacing its own worn-out parts while I do something else. At present, this ability to maintain and regenerate marks an absolute difference between living things and non-living machines. Even if we one day succeed in constructing a machine that can maintain and regenerate itself from within, the distinction may still hold, for it may be quite rational to regard any machine that complex, capable, and independent, as being alive.

Developing the structure of the body, which has occupied all of the foregoing chapters, is therefore a prelude to years of maintenance. Since maintenance means making more of the stuff made by development, it is natural to ask whether the mechanisms of repair simply recapitulate development or whether they are fundamentally different. The question is not just of intellectual interest—a clear answer is critical to any hope we have of improving our ability to repair damaged bodies and, perhaps one day, to try to combat ageing itself.

---

* This reaction, in a mix of cerium sulphate, citric acid, propanedioic acid, sulphuric acid, and potassium bromate, causes the oxidation state of the cerium to oscillate as it is reduced by propaedioic acid and re-oxidized by bromate. Feedback in the system causes slowly moving zones of these different oxidation states to form in a dish (they are visible because one of the oxidation states is yellow and the other colourless).

In the case of man-made machines, replacement of worn-out components can take place at almost any scale. Sometimes, very small components are replaced on their own, while at other times a complete large assembly containing hundreds of separate parts may be removed and replaced with a new version. On my ancient Land Rover, I have replaced individual items as small as lock washers, I have done new-for-old swaps of medium-sized, multi-component modules such as alternator and fan motor and, if the vehicle persists in its maddening habit of disengaging third gear at random moments, I shall probably replace the complete gearbox, with many hundreds of individual internal parts, because this will be much easier than taking the thing apart to deal with one sloppy shaft. The new gearbox will have been made in exactly the same way as the original was when the car was first built. Mammalian bodies do not maintain themselves in this way. Within cells, damaged proteins are replaced by new ones, and within tissues, damaged cells are replaced by new ones, but that is the limit of scale of replacement parts. Whole tissues and organs are not exchanged as complete units (except by transplant surgeons); instead, the existing organ is maintained continuously by very small-scale repairs from within. There are probably three main reasons for this. The first is that many organs arise, in embryonic and foetal life, from tissues that do not exist in the adult body. The endodermal lining of the yolk sac from which the gut first forms, and the somites from which vertebrae, muscles, and the inner part of the skin first form, are two examples of structures that have only a brief existence in embryonic life. They are simply no longer there to make new versions of these organs and tissues in the adult. The second reason is that many parts of the body subject to rapid wear and tear, such as the surface layers of the skin and the gut lining, last only a week or so before they have to be replaced, but it takes much more than a week to make these organs in foetal life. A complete organ new-for-old plan therefore could not keep up with demand. The third reason is the geometric and logistical difficulty of swapping new organ systems for old inside a crowded body. Maintenance therefore has to proceed by mechanisms that are quite distinct from embryonic development.

In principle, one could imagine a simple method in which worn-out cells would be replaced by proliferation of identical neighbours, something that we can call 'peer replacement' for the purposes of this discussion. This might involve some moderate complications, such as cells having to step back from their fully mature states that might involve very complex shapes and specialized metabolism, in order to be able to proliferate. It might nevertheless be reasonably straightforward. Each cell would, after all, have to replace only its own type, so the complicated mechanisms to decide which sort of cell to become, commonplace in embryonic development, would be unnecessary for maintenance. All that would be needed is a mechanism by which a cell can detect the fact that an identical neighbour needs to be replaced.

For all of its simplicity, maintenance only by peer replacement would run into serious problems in a long-lived animal. Many cells exist in a hostile environment where they are exposed constantly to agents that damage them. The lining of the intestine, for example, faces digestive juices dedicated specifically to digesting cellular components, while the outer layers of the skin are assaulted by dry air, ultraviolet radiation, wind, and bacteria. All cells in these locations are essentially in the same boat and, if one has accumulated so much damage that it dies, the chances are that its neighbour is by then also badly damaged. After a few generations of replacement of dead cells by proliferation of already damaged neighbours, there will be very little healthy tissue left.

There are two ways that an animal might deal with this problem without abandoning the idea of peer replacement. One would be to have a very short life and to avoid any stressful environments; it is quite possible that many very small animals have taken exactly this course (although, to the best of my knowledge, whether they use *only* peer replacement to replace worn out cells remains an open question). The other would be to invest heavily in damage control and repair mechanisms within cells. Such mechanisms do certainly exist: there is a set of enzymes and other proteins that can detect damage to DNA and repair it, membrane pumps exist that can expel small toxins from cells, and protein-destroying enzymes exist to make all cellular proteins short lived so that they are replaced rapidly with pristine ones. If a cell invested a great deal of energy and resources into these systems, it would remain healthier in a hostile environment, and peer-replacement would be more feasible. The problem is that the cost of such an investment in cellular repair processes to protect every single cell to this level, as would be necessary in a peer-replacement system, might simply be more than the resources (food) that the animal can procure. Even if it is not, by spending so much on cellular protection, the animal is probably leaving itself very meagre resources for growing and breeding.

A key feature of the problematic peer-replacement system is that all cells are equal and have to be protected equally, and therefore expensively. If an animal were instead to protect only a few of its cells, and to use only these to produce replacements for damaged ordinary cells, then it could economize considerably on its expenditure. If the highly protected cells could be located somewhere physically safer than an exposed surface, so much the better. Even better still, further savings in resources could be made if just a very few highly protected cells could make any of the cell types in the tissue. Such cells would lie effectively at the stem of a 'family tree' of the different cell types they can produce: cells that work this way are therefore called 'stem cells'. Maintenance of a tissue by the activities of stem cells offers considerable advantages over the peer-replacement system and, while peer-replacement may be used for quick patches to small wounds, stem cell-based tissue regeneration seems to be the way that large animals, including mice and humans, keep their tissues going over months and years.

One of the best-studied examples of maintenance of a tissue by stem cells is provided by the lining of the intestine. Here, the inner surface is exposed to a cocktail of digestive enzymes and bile salts and is physically battered by the passage of part-digested food. Even though they are protected, to some extent, by mucus, it is not surprising that most cells exposed on this surface do not last long. In mice, in which the cell biology of the intestines has been studied very carefully, most surface cells last only about five days; the figure for humans is probably similar. In a situation like this, it is unlikely that damaged cells could be replaced indefinitely, even in principle, by peer-replacement. Instead, they are replaced from a population of intestinal stem cells, tucked away in a comparatively safe niche. To understand how the intestinal stem cells do their job, it is necessary first to consider the anatomy of the intestinal wall.

The main function of the intestine is to absorb nutrients derived from food, and to re-absorb water that has been added to the food during digestion. Absorption takes place across the inner surface of the tube, and the total area of this surface is one factor that determines how much can be reabsorbed. Vertebrates have evolved two main methods to make this area as large as possible. One is to pack as long an intestinal tube as possible into the abdomen, by coiling it up rather than running it as a simple, straight connection from stomach to rectum. The other is to give the inner lining a complex structure of columns and ridges so that there is far more surface than there would be if the lining were smooth. In the small intestine, the inner surface is covered in small finger-like columns called villi that stick up to form the general level of the surface (Figure 79). Villi are present even at birth. Shortly afterwards, the regions between the villi fold downwards to make narrow depressions, called crypts. The bottom of each crypt is formed of Paneth cells, which specialize in secreting

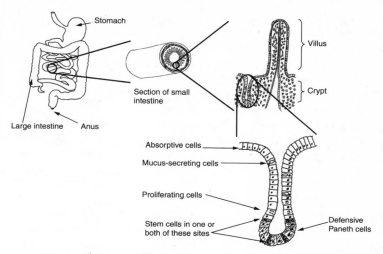

FIGURE 79 The anatomy of the intestine wall, and the structure of a typical crypt.

proteins that kill bacteria. The walls of the crypt above the Paneth cells contain many mucus-secreting cells. Protected by its position from mechanical trauma, protected by mucus from chemical attack, and protected by defensins from bacterial attack, the crypt is a much less dangerous environment than the villi. Unsurprisingly, it is deep in the crypts that intestinal stem cells reside.[2]

Somewhere, either between the Paneth cells right at the bottom of the crypt, or just above them, or possibly at both of these locations,[3] the intestinal stem cells can be found. There they proliferate, about once every four days. The daughters of this proliferation make a decision between becoming a new stem cell or leaving the stem cell niche and starting to move up the crypt walls. It is not yet completely clear whether the position of the new cell determines its decision, or the other way round: either way, the population of stem cells maintains itself and releases daughter cells that move up the crypt. As they move, these cells continue to proliferate so that the daughter of a single stem cell division has become up to sixty-four individual cells over the course of about three days. During this process, different cells become committed to becoming different types of intestine cell; some will become absorptive cells, some mucus-producing cells, some Paneth cells and some will become rare, specialized cells that make hormones. They keep moving upwards because they are effectively pushed up by new generations of cells that are forming behind them. The movement of the cells is not entirely passive, though, because those few that become committed to making new Paneth cells are able to move in the opposite direction and to head for the bottom of the crypt. They seem to be guided there by the same type of EPH/EPHRIN signalling[4] that was also used to guide the wiring of the eye to the brain (Chapter 13). By the time cells spill out of the top of the crypt, a few days after the original stem cell division that gave rise to them, they have matured enough to take part fully in the nutrition-gathering activities of the gut, and to replace cells that have died. Within a few days more, some will have been pushed right up into the villi where they will stay until they are so damaged that they too die, to be replaced in their turn by more cells newly minted in the crypt.

The rates at which intestinal cells are lost will vary with the health and diet of the person concerned. Someone who is eating but little, and soft food at that, may show relatively low levels of intestinal cell loss. Someone eating large quantities of rough, fibrous food, someone with a serious enteric infection, or someone whose food has been contaminated by a toxin, will show much greater levels of intestinal cell loss. Clearly, the stem cells and their proliferating daughters must sense how rapidly they are required to divide, because dividing too slowly to replace lost cells would result in a failure to maintain the intestine wall, while dividing too fast would risk blockage of the intestine with a mass of unwanted cells.

The mechanisms by which intestinal stem cells sense how rapidly to divide are still very incompletely understood, but it is fairly clear that signalling by WNT proteins, which played starring roles in several of the embryological chapters earlier in this book, is again involved.[5] The immediate daughters of the stem cells, and probably the stem cells themselves, show strong evidence that they are receiving a WNT signal from nearby Paneth cells and from more distant sources.[6] Furthermore, mice genetically engineered so that these cells cannot respond to a WNT signal show a complete failure of the stem cells to proliferate, and a failure of the intestine wall to be maintained.[7] Mice carrying the opposite type of mutation, so that they show a WNT response even if there is not actually any WNT around, show the opposite behaviour; and the cells proliferate far too much.

The cells surrounding the stem cells are the source of the WNT signal and, given that the proliferation of the stem cells seems to be controlled by WNT signalling, there must presumably be some link between the amount of the signal that is available for the stem cells to detect and the amount of repair that needs to take place. At present, though, the mechanism remains a mystery. Also mysterious is the mechanism by which the proliferating cells moving up the crypt make the correct judgement about which mature cell type to become, so that the ratios of absorptive cells, defensive cells, and mucus-producing cells remain correct. One possibility is that there is a self-organizing system in which each mature cell type secretes trace amounts of a signal suggesting that cells facing a decision become anything *except* that type of mature cell. In such a system, if one cell type is over-represented, its cumulative signal will be strong and will ensure that up-coming young cells become something else. On the other hand, if there is a shortage of a particular cell type, there will be very little signal saying 'don't make more of me', and the up-coming cells would therefore tend to choose to become that cell type and restore its numbers. All of this remains conjecture, and it will be interesting to see what the experiments say when they are finally done.

Another area of the body that is subject to a moderately harsh environment is the cornea, the tough outer surface at the front of the eye. As well as providing a protective outer covering for the eye, the cornea acts as a lens to help focus light on to the retina at the back of the eye. In fact, the cornea provides two-thirds of the lensing action of the eye, and the component of the eye that is actually called the lens provides only a third.

The cornea is irradiated by ultraviolet light, particularly in people foolish enough to go out into sunlight without protective sunglasses, it is scratched by grit and pollen, and it is swept several times a minute by the eyelids during blinking. It may also be attacked by cigarette or pipe smoke, particularly if the eye belongs to the smoker himself. Because it has to be transparent, the cornea also lacks a rich blood supply, which is important to the health of most tissues.

In embryonic life, the cornea forms from the ectoderm that directly overlies the lens as the eye begins to form. This is again a one-off event; once the cornea forms it replaces the ectoderm from which it came, so there is no chance of making a new cornea the way the original was made. Instead, worn corneal cells are produced from stem cells. As in the intestine, the stem cells are located in a more protective place than the danger-filled zone they service, and exist in a ring around the edge of the cornea, called the limbus (Figure 80).[8] The population of stem cells proliferates to sustain itself and also to give rise to daughter cells that are destined to make cornea and, as in the intestine, these daughter cells proliferate relatively quickly so that many corneal cells are formed from

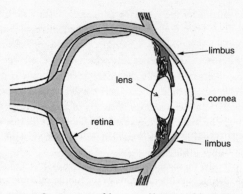

FIGURE 80 Anatomy of the eye, and location of the limbus.

each original stem cell division. Proliferating cells move away from the stem cell zone in the eye, radially inwards towards the very centre of the pupil, which is the place that is furthest from the stem cells. This movement from the limbus towards the centre of the eye has been illustrated dramatically by chimaeric mice.[9] Named after Chimaera, a mythical creature made from parts of more than one animal, chimaeric mice are produced from mixtures of embryonic cells, some of which are normal and some of which carry an experimentally introduced 'marker' gene that can be detected either when the animal is alive, or by a staining procedure after its death. In a chimaeric mouse, some stem cells carry the marker gene and others will not. When the eyes of an adult mouse are examined for expression of the marker gene,[10] they show a series of stripes, like spokes of a wheel, meeting at the centre of the cornea. Although these stripes look as if they are radiating away from the centre, they are in fact produced by cells coming in from the limbus, as examination of younger mice, in which the stripes have no yet reached the centre, reveals.

The intestine and the cornea are examples of hostile environments in which there is much attrition of mature cells and relatively fast proliferation of the stem cells. Many tissues are in much safer environments and their mature cells can last months or years. The stem cells that serve such tissues proliferate only very occasionally, but they can still be very important if large-scale repairs have to be made following infection or injury. An example of such a population of stem cells is provided by the kidney.

In kidney tubules, cells tend to be long lived in a healthy body, but they can be badly damaged and lost as a result of infection or poisoning. In development of the kidney, covered extensively in Chapter 10, the long, complicated tubules of the nephrons formed from groups of cells that aggregated together to make a cyst-like sphere which then elongated and bent to form the tube. This process, making new tubules from cells that are not initially parts of nephrons, is restricted to foetal development. Once humans are born, they can make no new tubules from scratch, but it seems that they can replace tubule cells from a tiny population of stem cells that exists in a specialized zone of each tubule, between the filter and the tubule proper (Figure 81).[11]

In the healthy body, these stem cells show very low levels of proliferation. When they sense damage to either the filter or the tubule, by some sensing mechanism not yet identified, the stem cells proliferate and their daughters move out along the tubule or into the filter, losing stem cell character and gaining the characteristics of mature cells as they do. There, they fill the gaps left by damage.

One system in which at least some of the signals are now known is the population of stem cells that maintains the blood. Circulating blood contains a number of mature cell types. Most common are red blood cells, loaded with the oxygen-carrying pigment, haemoglobin, that gives blood its characteristic

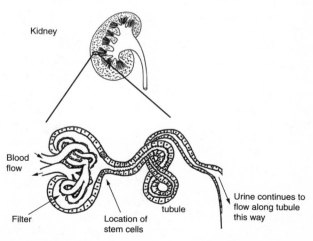

FIGURE 81 Location of stem cells in the tubules of the mature kidney (there may stem cells at other sites as well).

colour. Also present, at much lower numbers, are the various cells of the immune system, including phagocytes, B cells, and T cells (Chapter 17), and also small fragments of cells called platelets that are important in clotting blood at a wound. In early development, the first blood cells were produced by tissue interactions near the temporary kidneys (Chapter 9), but soon these tissues disappear and maintenance of the blood system produced switches to a stem-cell-based mechanism. This finds a temporary home in the embryonic liver but, once the long bones have developed, it moves to their centres where the mass of stem cells and developing blood cells forms a large proportion of the bone marrow.

As in the intestine, the stem cells themselves (haematopoietic stem cells, or HSCs) sit at the base of a branching tree of possible daughter cell fates (Figure 82). The population of HSCs proliferates slowly, maintaining its own numbers and giving rise to daughter cells that are committed to eventual maturation. These daughter cells proliferate strongly to create rapidly expanding colonies of cells, the members of which will later refine their choices of fate in the succession of steps shown in Figure 82. Again, the proliferation of each cell has to match itself to the body's needs and exactly make up for the blood

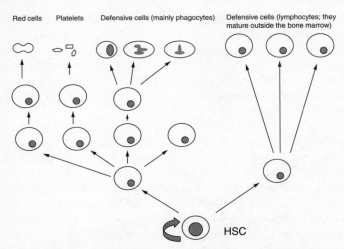

FIGURE 82 The 'family tree' of cells that are produced in the bone marrow, ultimately, by haematopoietc stem cells, marked 'HSC' in the stem of the tree. The population of these stem cells maintains itself and gives rise to the cells above it. The cell types above proliferate as well, resulting in very large numbers of final, mature cells being produced by each proliferation event at the level of the heamatopoetic stem cell at the bottom of the tree.

cells that die. If the marrow were to make too many new cells, the blood would make a dangerously thick soup, whereas if it made too few, the blood would not be able to carry enough oxygen or be useful in defending the body. The balance of cell types also has to be regulated carefully and can change according to circumstances, the number of defensive cells increasing during infections, for example.

Careful studies with cultured bone marrow cells has identified many signalling molecules that alter the extent to which marrow cells proliferate, each type of cell responding to a different set of these signals. Cells, that have already made the decision to contribute to the production of red blood cells rather than other types of blood cell, carry on their surfaces receptors for a hormone called erythropoietin. When levels of this hormone are low, the cells just sit around, proliferating little. When levels of erythropoietin are high, the cells proliferate vigorously and their daughters mature into new red blood cells. Erythropoietin itself is produced mainly by a part of the kidney where the supply of oxygen-carrying blood is, for anatomical reasons, relatively poor even in a healthy individual. The kidney cells sense the concentration of oxygen around them using the same molecular system described in Chapter 9, in the context of cells requesting new blood vessels. The lower the oxygen concentrations they sense, the more erythropoietin the kidney cells make and release. This erythropoietin travels to the bone marrow, where it stimulates cell proliferation and therefore production of new red blood cells. These endow the blood with more oxygen-carrying capacity, and the levels of tissue oxygen rise. Sensing this, the kidney cells reduce their production of erythropoietin and the system settles down so that red cell production is just enough to balance losses and maintain optimal numbers.

The overall system therefore organizes itself to produce just the right number of new red blood cells, without any of the components having to 'know' anything more than how to respond to a simple chemical signal. The mechanism works well provided that the only source of erythropoietin lies within the healthy tissues of the body, but it can be upset by disease states that create false signals. When the blood flow to a kidney is reduced, for example by a diseased or damaged renal artery, the cells of the kidney become very short of oxygen, even when the rest of the tissues of the body are fine. The kidney cells therefore produce very large amounts of erythropoietin, and too many new red blood cells are made as a result. Similarly, if extra erythropoietin is injected from outside the body, the bone marrow is fooled into behaving as if there were a crisis in red blood cell numbers and over-produces new red blood cells. The recent history of athletics has featured a number of scandals in which competitors have dosed themselves with erythropoietin to increase the oxygen-carrying capacity of their blood, and therefore the maximum power their muscles can sustain, in exactly this way.

The production of defensive cells and their precursors in the bone marrow is also regulated by signals from the body. Infection by microbes, for example, triggers existing defensive cells such as T cells and phagocytes (Chapter 17) to produce long-range signalling molecules. These spread, via the blood system, throughout the body and reach the bone marrow, where they stimulate cells to produce a cocktail of local signalling molecules. These local signals drive the proliferation of cells already fated to make more cells of the immune system, and also drive the rapid maturation of cells to become useful defenders. In this way, the bone marrow responds to a microbial threat somewhere else in the body by rapidly delivering reinforcements.

The variable stimulation of red blood cell production by erythropoietin, and defensive cell production by other signals, might solve the problem of maintaining adequate numbers of cells in the circulation, but it creates a problem in the bone marrow itself. The red blood progenitor cell population, whose proliferation and maturation is driven by erythropoietin, cannot maintain itself indefinitely and all of the cells will eventually be lost as they mature and leave the marrow as red blood cells. They have, therefore, to be replaced, by division of cells below them in the 'family tree' (Figure 82, and these in turn will run out and have to be replaced, ultimately by division of the HSC stem cells themselves. The problem of controlling proliferation therefore extends downwards to every level of the tree. It seems that it is solved by a relatively simple generic signalling mechanism that operates similarly for all of the cells, although different molecules are used for different cell types.[12] Basically, each cell secretes molecules that inhibit the proliferation of the cell type below it in the family tree. If there are enough cells at a higher level, they will together make enough inhibitory signal that the cells below them are quiet. If the population at a higher level falls, because cells there are maturing and leaving the bone marrow, there will be less inhibitory signal and lower levels therefore proliferate more and their daughters replenish the level above them. This system extends downwards, level by level, to the ultimate stem cells resident in the bone marrow, the HSCs themselves. Although there certainly are other signals in the bone marrow system, computer modelling strongly suggests that these downward-facing inhibitions might be the dominant regulators of proliferation.

The importance of the bone marrow as the location of stem cells that maintain the blood has been appreciated for decades, even if the details are still being worked out. More recent is the growing appreciation of the bone marrow as a location of stem cells that can maintain a range of distant solid tissues as well. Indeed, it is possible that the bone marrow can produce cells to maintain every type of connective tissue in the body, and this observation is driving a revolution in the way that we understand human biology. The stem cells behind this are called mesenchymal stem cells,[13] and current evidence suggests that they are themselves daughters of the HSCs described in the section about blood. When they are removed from the body and kept in laboratory culture, mesenchymal stem cells can proliferate and produce an astonishing variety of mature cells, including those of connective tissue, fat, cartilage, and bone. Different culture conditions favour the production of different types of cells.

Showing that bone-marrow-derived mesenchymal stem cells can produce these cell types in the laboratory, and showing that they really do so inside a body, are two different things. An unintentional 'experiment' done directly in humans suggests that these cells really do contribute to maintaining distant connective tissues. The lives of people whose own bone marrow is destroyed, typically by exposure to high doses of radiation or by aggressive chemotherapy against leukaemia, can be saved by providing them with a little healthy bone marrow from a donor. Provided that the donor and recipient are similar enough that their tissues do not reject each other, the donated cells settle in the marrow of the recipient's bones and there they start to make new blood. Matching donor and recipient so that they are similar enough involves making sure that their cells make the same types of cell surface proteins—a bit like the well-known system of A, B, and O for blood donation but rather more complicated. Whether or not donor and recipient are of the same sex is irrelevant, and many donations have been made between brother and sister or father and daughter.

All the cells of a boy or man contain a Y chromosome, but those of a girl or woman do not (Chapter 12). This means that when bone marrow from a male donor is grafted into a female recipient, the donor cells carry the Y chromosome as a detectable genetic feature that is absent from the cells of the recipient. Assuming the operation is a success, the recipient will continue to live, hopefully for many years, and her body will maintain itself. If mesenchymal stem cells in the bone marrow naturally contribute to maintaining the connective tissues of the body, then some of the specialized cells in those connective tissues would contain a Y chromosome that could be detected by microscopic examination tissues obtained by biopsy or autopsy. This turns out to be the case: the connective tissues of female recipients of male bone marrow donors contain some cells with a Y chromosome. Such cells have been described in the connective tissues of heart,[14] gut,[15, 16] brain[17, 18, 19] and kidney,[20] and in all cases they seem to be fully parts of the tissues concerned (and not, for example, just blood cells passing through). The Y chromosome even turns up in some tissues that are not connective at all, such as the tubules of liver and kidney, the inner surface of the intestine, and the neuronal cells of the brain. Bone-marrow-derived cells have also been observed contributing to pathological structures made by host cells, such as the cell masses of endometriosis (a benign growth of uterine lining in the wrong place) and to the cell mass of a malignant carcinoma. Emphasizing the point made in Chapter 12, that most cells of the body pay no attention to whether they carry a Y chromosome in deciding whether to make male or female structures but rather heed their environment and hormones, Y-chromosome-bearing donor cells have been found as fully integrated cells of a recipient's uterus.[21, 22]

The evidence above does not prove that mesenchymal stem cells are the source of each of these cell types, but it does prove that some sort of stem cell in the bone marrow must be. The same result is now being seen by other cell-marking techniques that involve no transplantation at all, but rather make use of spontaneous genetic changes that happen in the bone marrow of some people and therefore mark some bone marrow cells.

Intriguingly, the presence of Y-chromosome-bearing cells has also been detected in the tissues of women who have never received a transplant, but who have given birth to a male baby.[23, 24] Presumably, stem cells from the male embryo passed through the placenta and colonized the mother. They can be detected years later, and in one study[25] were present in every tissue that was examined: a child forever having a place in its mother's heart turns out to be more than just a metaphor.[26]

Accepting that the bone marrow can contribute to maintenance of distant tissues, it is still not clear how important this is compared to maintenance by stem cells resident in the tissues, cells that have nothing to do with bone marrow. The techniques used to detect bone marrow contribution are very sensitive and there is a chance that these experiments have identified a fascinating, but ultimately very minor, method of tissue maintenance. In the example of the intestine, for example, the contribution from bone marrow seems tiny compared to that from the stem cells in the crypt. Even if repairing tissues from the bone marrow does turn out to be a minor method of body maintenance, it may still be useful medically, as will be discussed further in the final chapter.

As we noted at the beginning of this chapter, stem cells generally take very good care of themselves. Many expend considerable energy and resources into making pumps and channels to expel toxins. They are also very sensitive to damaged DNA and kill themselves rather than struggling on with damage. Presumably, this state of affairs arose through natural selection, over evolutionary timescales, against the consequences of damaged cells proliferating. The care that stem cells take of themselves, and their tendency to kill themselves if too damaged, is very real, and has a dark side as far as human health is concerned. One example is provided by acute radiation sickness.[27] A few days after a moderate or severe exposure to ionizing radiation, a typical victim will begin to suffer severe, bloody diarrhoea, along with vomiting, hair loss, and opportunistic infections. One of the causes of the diarrhoea is the loss of many intestinal stem cells in the crypts, which have detected DNA damage and eliminated themselves: with so many gone, the intestinal lining, itself more damaged than usual, cannot be maintained.[28] People who survive this stage often die within weeks, due to a similar loss of bone marrow stem cells, and thus the ability to renew blood cells (including those that maintain body defence). The general cells of their bodies, although somewhat damaged, may well not be in too bad a way, but the intolerance of the stem cells for this amount of damage can mean a lingering death, as it has for the victims of accidents and deliberate actions, such as those at Los Alamos, Hiroshima, Nagasaki, Bikini, Kyshtym, Vinča, K-19, K-8, K-431, and Chernobyl.

The damage-detecting systems of stem cells are not perfect, and occasionally even stem cells will suffer mutations. Many will be 'silent', that is, of no consequence to the behaviour of the cell, but a few are much more significant. We have already considered the importance of the WNT signalling system in controlling the rate of proliferation of intestinal stem cells and their daughter cells, and have noted that mouse mutations that force WNT signalling 'on' in these cells make them proliferate too much. Intestinal tumours are the third most-common cancers in adult humans, and most show activating mutations in the WNT signalling system.[29] By the time the cells have made an aggressive, spreading cancer, there are other mutations as well, but the frequency with which the WNT mutations turn up suggests that they may be fundamental to this type of cancer.

For at least some types of colon cancer, the intestine's normal pattern of having stem cells that renew their own population and produce proliferative daughter cells remains even when the cells have altered to become a tumour.[30] The stem cells now proliferate outside normal control and give rise to daughter cells that do the same, but while the mutant stem cells can maintain their own population, their daughter cells eventually, after many divisions, 'run out of steam' and die, being replenished by daughters of later stem cell divisions. Similarly, the cancer stem cells can found a new cancer in a host animal but the other cells of the tumour cannot.[†] The normal structure of the intestine wall is lost in the tumour, so what the mutant stem cells and their daughters are now doing makes no anatomical sense, but the basic logic is still there. Also still there is the stem cells' habit of looking after themselves,[31] Cancers that cannot simply be removed surgically are typically treated with chemotherapy, that is with small drugs that are particularly toxic to dividing cells. Stem cells can be very good at expelling these drugs to protect themselves, and also good at repairing the damage the drugs do. Furthermore, although stem cells divide, they often do so much more slowly than the daughters that are (in normal life) destined to move out of the crypt. In a normal human intestine, stem cells divide about once every four days whereas their daughters that move up in the crypt divide about once every twelve hours. The stem cells are therefore less vulnerable to the effects of drugs aimed at dividing cells as well as being better at expelling them. There is therefore a real danger of the drugs killing all of the tumour *except* for the stem cells which, given their unique ability to found new tumours, were the most important cells to kill. In this view of cancer, relapse of a tumour after a few years of apparent freedom of disease is due to the survival of occasional stem cells.

It must be stressed that, while there is very good evidence for cancer stem cells in certain tumours, cancer biologists are by no means agreed about whether this is a generally valid model for all cancers, or even for most.[32, 33] It is an urgent research question, because an answer would help inform the strategy for developing future treatments, perhaps ones aimed specifically at the stem cells themselves. Given that each side of the argument seems to have a solid body of evidence, it may simply turn out that some cancers are based on cancer stem cells and some are not, and that working out which type a patient has may become an oncologist's most immediate concern.

[†] This experiment required the host animals to have been treated so that their immune systems could not react to the foreign tumour. Human cancers cannot pass from one person to another person.

Cancers are not the only diseases that involve misbehaviour of stem cells resident in tissues. Too much proliferation of stem cells and their daughters, even without cancerous change, can cause serious problems in tissues. Following hard on the heels of the discovery of the kidney tubule stem cells described earlier in this chapter was the realization that the main problem in a serious kidney disease seems to arise directly from an error in stem cell behaviour.[34] The main pathological feature of the disease crescentic glomerulonephritis is that the filtration units at the ends of the tubules become damaged and replaced by a crescent-shaped mass of unspecialized cells, useless as a filter. If this happens in too many kidney tubules, it can lead to kidney failure. Judging from the proteins they express, the cells of these crescents may be the daughters of the tubule stem cells that have proliferated far too much to produce a mass instead of turning into replacement filtration cells. The over-proliferation is nothing to do with cancer, and probably results instead from some underlying cause of the disease causing the kidney to make signals calling for proliferation when none is in fact needed.

Failure of stem cells to proliferate enough leads to collapse of the tissue that is supposed to be maintained. In the eye, for example, the disease aniridia-related keratopathy, which results in the transparent cornea being replaced by a milky-white, scar-like tissue, seems to arise from a failure of limbal stem cells to maintain the cornea, although this may not be the only problem.[35] The result is blindness. In this particular disease, the problem seems not to be with the stem cells themselves as much as a defective limbal environment, which prevents the stem cells from behaving properly.

The stem cell failures described above affect only a relatively few unfortunate people. One major failure of stem-cell-based maintenance affects us all, however: the replacement of damaged and worn-out tissue is imperfect, and as a result we slowly accumulate damage. In other words, we age. There are many different theories about why we age—something that small and simple organisms tend not to do. One explanation is that we simply accumulate random damage, from radiation, free radicals, poisons, etc., at a faster rate than can be repaired or diluted by cell division. Even in a body in which many stem cells are still pristine, gradual degradation of the bulk tissues creates problems for their maintenance. Damaged cells make misleading signals that confuse maintenance processes, proteins between cells can become cross-linked and difficult to clear out of the way, and problems like these can make repairs imperfect. Slowly but inexorably, these problems stack up, minor at first but accumulating more and more quickly the more abnormalities are around to inhibit natural repair. Thus ageing becomes faster as time goes on. Once it is severe enough to seriously affect the working of one of the body's essential systems, the kidney perhaps, or the heart, the whole internal environment begins to become abnormal and harder to repair, and the rate of decline even steeper.

In principle, it may be possible for bodies to invest more in repair and to age more slowly. Indeed, a growing number of genetic experiments has resulted in organisms that have been engineered to age significantly more slowly than normal, although they do still ultimately age and die. To understand why we, and other organisms, do not already carry sets of genes that are already optimized for lives as long as possible, it is necessary to consider how evolution works.

Consider a starting population of animals that vary, from individual to individual, in the balance of the resources they put into repair for a long life versus living energetically and briefly. The proportion of each type of animal in the next generation will depend on how many offspring of each type of animal survived to maturity. Potentially long-lived animals have the chance of reproducing many times, which is to their advantage, but this depends on their ability to find mates, food, and territory. Rivals that put all of their resources into energetic and short lives, with little spent on maintenance, will have fewer total years for breeding but, if their energy allows them to find mates, food, and territory more efficiently, they may still do well. In a situation as simple as this, long life may be the option that yields the largest number of offspring and so the one that ends up dominating the species. If the action of predators and diseases is added, though, the balance can change rapidly. If there is a high probability of being killed by something on any given day, then the investment in repair mechanisms that allow a long life becomes a lot less advantageous, as much of the investment is likely to be wasted. There is no point in an animal having repair systems that allow it to live a century if it has a 50 per cent chance of being eaten every year. Under these circumstances, the alternative investment, in fast living and rapid reproduction with little thought to the morrow, shows real advantages.

This theory, that predation risk encourages the selection of gene sets that do not put large resources into longevity, gains some support from comparisons of similar animals that are subject to mild or severe pressure from predators. Little brown bats, for example, have few predators and invest significantly in long-term maintenance. They live for around thirty years in the wild. Mice, which are about the same size, suffer a very high rate of predation. They reproduce rapidly but invest little in maintenance, so that even when they are kept as pets and therefore completely safe from predators, they tend to live only three years, a tenth of the normal life of the bat.

The fossil record of humans points to our species, *Homo sapiens*, first becoming distinct from other hominids only a few hundred thousand years ago, in predator-rich Africa. Although our ancestors had used tools for several million years before then, it is a reasonable assumption that they were still at a risk of predation not too different from the great apes, at least until the sudden burst of technology that began with the Neolithic period ten thousand years ago. Ten thousand years—just five hundred generations—is a short time in evolutionary terms and we probably therefore still carry the compromise between youthful vigour and investment in longevity that was selected by the predation our ancestors suffered on the African plain. Even now that the risks of disease and predation have largely disappeared for those of us lucky enough to live in the first world, there is no particular selection pressure for long life. Indeed, the fact that an orphaned child will be looked after to maturity even by non-relatives removes one of the few evolutionary pressures that couples the long life of a parent to the success of its offspring. In terms of our natural constitutions, we are probably therefore stuck with repair systems that leave very few people living more than a century. If we want better repair systems, so that we can live longer, we will have to do something about it ourselves, putting to practical use what we are learning about normal development and repair.

PART IV

PERSPECTIVES

19

PERSPECTIVES

γνῶθι σεαυτόν *   Socrates

The preceding chapters look back on a remarkable journey that every one of us has made, a journey that took us from a simple, single-celled existence to the organized collective of ten trillion cells that constitutes an adult human being. The complete enterprise involves far more events and processes than can be described in a short book, and probably far more than have yet been discovered. There is no need, however, to know the details of every single event in embryonic development in order to begin to understand how it works. A major part of the 'art' of science is the skill to perceive general truths from a limited number of examples, and the history of science is replete with instances of this: the movements of just a few planets led to Kepler's Laws and thence to Newton's Law of Universal Gravitation, the layering of a few rock formations in Scotland led Hutton to formulate the first modern geological theory for the whole Earth, and the variations between the relatively few species he studied led Darwin to a theory of evolution by natural selection that applies across all life. Even from the limited number of developmental events that have been studied, and the small fraction of these that have been presented in this book, it should be possible to gain some broad understanding and perspective on how a human body can organize itself from almost nothing.

* Know thyself.

One very strong theme that has been at the core of almost every event discussed is communication between cells. At every stage of development, protein-based machines detect signals from a cell's environment, be these signals mechanical (tension, free surface) or biochemical (molecules from other tissues), and a combination of these signals and the existing internal state of each cell determines what it will do next. This rich communication between components is quite unlike conventional engineering, in which relays and transistors pay no attention to one another in the construction phase even if—as in a computer—they communicate extensively when the completed machine is switched on. Because rich communication between components is an aspect of life that is so different from the inanimate world, it may be a good place to start when seeking a model for embryonic development that is better that the ones we have already rejected.

In the examples set out in previous chapters, signalling between cells achieves two things: it increases complexity, and it corrects errors. Biological complexity is hard to define in a precise numerical way,[†] but for an embryo it can be taken as a measure that includes the number of distinct cell types and anatomically distinct structures present, ignoring those structures that are inside cells.[1] By this measure, it begins low (one type of cell, one structure) and becomes high (hundreds of cell types, thousands of internal structures). What is more, during the most active phase of development, the rise in numbers of structures is approximately exponential (Figure 83).

Exponential growth is characteristic of systems in which already-achieved growth bequeaths an increased ability to grow. The classic example is the population size of bacteria in liquid culture. One bacterium grows and divides into two, and each of these can now grow and divide to yield a total of four, which go on to make a total of eight, then sixteen, and so on, the net increases getting larger each time. Exponential rise in complexity suggests a similar effect, in

---

[†] In less messy fields of science, complexity can be measured as the minimum amount of information needed to specify an object exactly. A simple run of identical digits, '1111111', is therefore less complex than an equally long run of random digits (e.g. '1576249'), because the first can be stated as just '7 1s' while the random sequence must be spelt out fully; similarly, the shape of a sphere is simpler than the shape of a rock. Trying to measure biological complexity in exactly this way is not only difficult (are we talking of complexity of shape, or of cell states, or what?); it also risks a dangerously circular argument. Assuming that all of the complexity of the body arises from the genome alone in a blueprint-like way *could* yield a number (the size of the genome), but this is uselessly circular if one is trying to use that number in a discussion about how complexity really arises and its real relationship to genes. It would also ignore all of the information in the proteins of the egg that control the genome in the first place (Chapter 1).

Mouse data

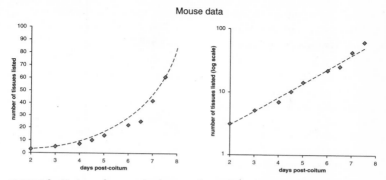

FIGURE 83 During early mouse development, the number of distinct tissues increases exponentially with time. The two graphs show the same data, the left graph being a linear plot of the data with an example exponential curve, and the right graph being to a log scale, in which exponential relationships appear as straight lines. As development proceeds beyond these stages, the rate of increase of tissue types starts to flatten off. The tissues plotted are those listed in the Edinburgh Mouse Atlas Project, <www.emouseatlas.org> (accessed 6 July 2013).

which complexity achieved increases the capability of an embryo to add more complexity in the next step. We have already seen how communication can achieve exactly this. Once a difference exists between cells within an embryo, cells can use detection of this difference to become a third cell type. This creates two new boundaries between cell types, and each of these can be used to perform the same trick. Figure 84 depicts this schematically for a simple row of cells in a tissue, and Chapter 7 illustrated a real example, describing how differences in signals coming from the ectoderm and notochord are used to generate complex patterns of cell types in the somites and the neural tube. In real life, as in the simple diagram, differences feed on differences: the embryo uses this effect to bootstrap itself from dull uniformity to exquisite internal diversity and organization. As recounted in Chapter 3, it seems to use a simple physical difference as a seed to start the whole process off.

The second powerful use of cell communication is in balancing the amounts of different tissues and correcting errors made inevitable from the random thermal noise inherent in biochemical reactions. The way in which one tissue can request the growth of another to serve it, as when a hypoxic tissue requests a blood supply (Chapter 9), the way in which the size of a cell population depends on the body it has to serve (Chapter 16), the way in which cells that are in the wrong place elect to kill themselves (Chapter 14), and the way that the

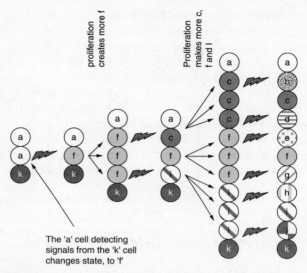

proliferation creates more f

Proliferation makes more c, f and l

The 'a' cell detecting signals from the 'k' cell changes state, to 'f'

FIGURE 84 Boundaries between two types of cells can cause cells at the boundary to become a third cell type. This creates two new boundaries, which can be used to specify yet more cell types. This schematic is meant to show the general idea, but not any specific part of the embryo. Chapter 7 described real examples of this type of thing, in the neural tube and somites.

proliferation of stem cells is controlled by signals from the cell types they themselves make (Chapters 16 and 18), all illustrate how very flexible our development is.

At the heart of this flexibility are signalling loops: not just signals, but *feedback loops*. In these loops, the result of a process is fed back to control the process itself. An example was presented in Chapter 9, where the oxygen brought as a result of capillary growth reduces production of VEGF signals that would otherwise drive further capillary growth. The presence of feedback loops gives the communication between cells the character of a true conversation, signals being answered by other signals coming back, directly or indirectly, so that the behaviours of cells are strongly inter-dependent. This is probably the key to how biological construction can take place without an external builder reading plans and comparing them to what he has so far completed. House-bricks have no way of sensing the state of the building project of which they are a part, and no way of altering what they do according to what they sense. Cells do; they cannot 'see' the whole embryo as a builder can stand back and look at his handiwork, but they can sense all they need to sense to behave in the right way.

The idea that constant communication between components can replace the need for an external builder or organizer is not unfamiliar from everyday life. An extraterrestrial onlooker who knew nothing of being human might assume, looking at the movements of a crowd of shoppers in London's Oxford Street, or of couples dancing around a busy ballroom, or of a throng of music fans at an open air concert, that there must be some master choreographer organizing everything to avoid anyone colliding or being crushed to death. Having participated in such things, we know that each human is simply responding to local cues around: although the very nature of the environment denies everyone any kind of global picture, the great mass of humans organizes itself in a safe and sensible way.[‡] At a larger scale, the core mechanisms of civilization, such as the development of language, the mechanisms for an economy, the means of food distribution, and even the means of scientific enquiry, have generally emerged from the interactions of a great number of people, none of whom had the global perspective that would be the prerogative of an outside observer but instead behaved according to a very limited and fairly local knowledge. Nevertheless, civilization seems to be reasonably well organized and relatively robust, even acknowledging occasional economic upheavals. Indeed, experience from the limited number of occasions in which a particular human has tried to plan a language or an economy suggests that this is less effective than the traditional pattern of self-organization by masses of people, each of whom acts according to fairly local circumstances. While constructing multi-individual organizations, from families to societies, we communicate intensely, as do termites, ants, and bees, which make social colonies. However imprecise, this analogy can be helpful in understanding how communicative cells can also create organizations much larger than themselves.

A striking feature of developmental mechanisms is the way they are nested, later ones including earlier ones to run fine-scaled events. Building a working sensory nervous system depends on adjusting the strengths of neural connections, which uses positive feedback learning loops (Chapter 15). These depend

---

[‡] Usually: there have been infamous exceptions, and an interesting area of academic research now explores crowd behaviour with a view to making environments that discourage dangerous departures from the normal, safe processes of self-organization with which we are familiar.

on achieving an abundance of connections in the first place, which involves migration of neural crest cells and of their processes according to interpretation of navigational cues (Chapter 13). The ability to couple navigation to cell movement depends on the self-organizing feedback loops of the cells' leading edge (Chapter 8), and these loops depend in their turn on the simple, physical self-assembly reactions of protein units that come together to form microfilaments (Chapter 1). The nesting may even extend upwards, beyond the level of an individual body, to shape human societies. Care should be taken not to take the analogy too far, however: societal organization has added very significant new capabilities, such as transmission of acquired knowledge down generations, that have no obvious counterpart within the body, and bodies have features, such as reproduction being restricted to the germ line, that have no obvious counterparts in most human societies (but some social ant and bee colonies do restrict reproduction to a few individuals, which are effectively the colonies' germ lines).

How does the communication-centred view of human development outlined above fit with the gene-centred view that is currently so prevalent? The two views can be brought together simply by noting that the protein machines that lie at the heart of the communication-centred view are themselves constructed according to information in the genes (Chapter 1). Proteins regulate the expression of genes, and genes specify the production of proteins; there is no conflict between a properly articulated gene-centred view and a communication-centred one, because they are two sides of the same coin. There is, though, a conflict with a way that a gene-centred view is often expressed. This view seems to be based on a misunderstanding of a laboratory shorthand used by scientists, which has escaped the walls of the lab to cause much general confusion, even in biology students, and which should caution us all to be careful about how we express ourselves, even in the privacy of a white-benched sanctum. The shorthand in question has the form 'the gene for X', and it has encouraged a view that there are specific genes that directly specify particular high-level attributes of the body, such as long noses, strong arms, or a high IQ.

Classical genetics is mainly a study of correlations, specifically the correlation between the mutation of a gene and a detectable effect on the organism. In a developmental context, discovery of a correlation means that a gene can be identified as 'a gene, this mutation of which means that body part X does not form normally'. This is too much of a mouthful to keep saying in a research meeting, and it quickly becomes reduced to 'the gene for X'. When the gene is finally named, its name may well carry the flavour of this shorthand: *wingless*, *thick vein* and *small eye* are examples. As long as everyone remembers what this shorthand really stands for, there is no problem. If they do not, though, the phrase can all too easily suggest that the function of a gene is to make a specific body part, and further it can suggest that there is a 1:1 relationship between genes and body parts, something that has become a very common misconception. The problem comes from the directional nature of causality, something that is neatly encapsulated in a famous English nursery rhyme (probably based on events that cost the English King, Richard III his life), that runs;

> For want of a nail the shoe was lost.
> For want of a shoe the horse was lost.
> For want of a horse the rider was lost.
> For want of a rider the battle was lost.
> For want of a battle the kingdom was lost,
> All for the want of a horseshoe nail.

Here there is a clear causative correlation between absence of a particular component and a dramatic outcome. No reasonable person, though, would consider the primary function of a nail to be winning the battle of Bosworth Field. The function of the nail is to hold on a horse's shoe, and a teacher describing the object as 'the nail for Tudor rule' would give students a very peculiar view of history. In a similar vein, it is misleading to consider the function of the unmutated version of the fruitfly gene *wingless* being to make wings. It is not: the function of the gene is to make a signalling protein, the fly equivalent of the human WNT proteins described earlier in this book. The signalling protein has multiple functions in the fly, one of which is important in making wings. In certain types of mutation of the gene, the rest of the fly is still able to develop fairly normally but the wings do not, which is how the gene was first identified.

It may seem that the distinction is subtle, and that caring about whether people take literally the phrase 'the gene for X' is no more than academic pedantry, but I would argue that it matters a great deal. The problem with the 'Gene for X' view is that it evokes the idea of a fixed plan, rather than a multi-layered nesting of mechanisms that organize the body by integrating their own signals and those from their environment. It therefore gives a quite false impression of determinism. The pervasiveness of this misunderstanding is illustrated by the vast amount of hot air generated in angry debates on the balance, in shaping a human life, between genetics and environment, between 'nature' and 'nurture'.

As early as 1909, the biologists Wilhelm Johannsen[2] and Richard Woltereck[3] published, separately, two clear sets of evidence that the development of animals results not from their genes alone, but from the interaction of the mechanisms specified by the genes with their environments. Over the following years, many other studies supported this in a vast number of species including humans. Despite this, over-literal readings of shorthand lab phrases such as 'gene for X' created, in the minds of some psychologists, sociologists, educationalists, politicians, and the public, a quite false dichotomy between nature and nurture. When politicians and physicians use misunderstood science as a basis for designing education systems, mental care systems and penal codes, the issue becomes very important indeed.

Properly used, information about faulty development and its causes, be they genetic or environmental, has been and continues to be of massive importance to the understanding of normal development. Teratology, the study of abnormal development, was named from the ancient Greek for 'monster' or 'marvel', although most people switch to value-neutral terms when discussing unusual humans. Teratology makes two broad sets of contributions. The first is association of a particular gene or chemical pathway with a particular developmental event. An inactivating mutation in both copies of the gene encoding the signalling molecule GDNF, for example, results in the birth of a mouse or human that has no nervous system in its intestine, and also in the absence of kidneys. This observation immediately suggested that signalling by GDNF must be involved in the development of both of these organs, something that was quickly verified using genetically normal embryos and direct, local manipulation of GDNF signalling using artificial sources or inhibitors. This way of thinking has been so valuable that massive programmes have been run to determine the effects of mutating (independently) essentially every gene in simple organisms such as worms and flies, to gain some idea of what kinds of developmental events require the proteins encoded by each gene. Very often, it has been possible to apply the information from these simple animals directly to experimental effort on mouse and human embryos, with great success. Similarly, close association of chemical toxin with a specific developmental failure can, where the biological action of that toxin is understood, be used to identify signalling pathways and other processes that are required for the development of some body part. The case of thalidomide was described in Chapter 11. Another example, originating in agricultural rather than experimental animals, is the frequent production of lambs with just one central eye and one nostril, in flocks grazing on land that harbours the corn lily. This plant contains cyclopamine (named from 'Cyclops', the one-eyed primordial giant of Greek mythology), which turns out to be a powerful inhibitor of the Hedgehog signalling pathway that has been mentioned several times in this book. Again, observations like this quickly point researchers to the probable involvement of Hedgehog signalling in correct patterning of the face. Much of our understanding of human development presented in this book has been obtained with the aid of teratological (including genetic) reasoning.

The second contribution of teratology, again in its broad sense that includes genetics, is in understanding the other question of how we humans came to be, not the question in the developmental time that has been the focus of this book, but the other version of the question that applies to geological time. As Wallace and Darwin realized, evolution requires two things—variation, to generate a mixed population, and natural selection to bias which variants of that population are best represented in the next generation. Variation in the animals competing to leave successful offspring implies variation in the precise developmental events that made them. Most of the time, these may be small—a leg a little longer or shorter, a lung a little more branched, a cerebral cortex a little more folded. Sometimes, the effect of a mutation may be much larger—a mutation that causes a jump, for example to eliminate elective cell death in foot webs to create webbed feet well adapted for swimming. Evolutionary biologists still debate whether most evolutionary change operates at the level of frequent very small changes or at the level of sudden jumps but, in either case, variations in development provide the variation in mature organisms that will leave behind their progeny. Study of how changes in an animal's genes can result in changes in its body (some large, some small), and also how environmental influences ranging from temperature to toxins can interact with genetics to alter development, illustrates how evolutionarily important variation might arise.

There is much about human development that we do not yet know. Are we likely to discover any major surprises that revolutionize our view of what is going on, or have we now unearthed the main principles by which development works, future work just being a filling-in of details? The general assumption of scientists in the field seems to be that we probably do now know the general principles—gene control, cell communication, cell movement, etc.—but the history of science warns us that such consensus is no guide to the actual state of affairs. Late Victorian physicists were mostly confident that, with Newton's Laws, Maxwell's Laws, thermodynamics, and a few other things, they knew the basic mechanisms of the universe and all that remained was filling in detail. Then relativity and quantum mechanics were discovered and physics was shaken to its core. Nature keeps surprising us. If there are any truly paradigm-changing discoveries to be made in developmental biology, where might they be found?

It seems unlikely, at least to this author, that further studies at the molecular level will yield truly revolutionary findings. Such studies certainly retain their ability to come up with some surprises, as the recent discovery of RNA interference and micro-RNAs,[4, 5, 6] a previously-unsuspected method of gene control, confirmed, but even this did not alter the basic idea that genes control the expression of other genes; it just said that sometimes this can be done using RNA rather than protein. A more promising way to study genes is by looking at overall patterns of gene expression, noticing groups of genes that always seem to act together and that may therefore act as a coherent system or 'module' to perform some important function. A related place to look for significant new truths might be in the patterns of connection in the communication networks that cells use: not the molecular details of the signals, but the overall patterns—the 'wiring diagrams'. So far, signals tend to have been studied one by one, but a few pioneers have recently been looking at complete networks of signals to see if they can see any patterns. In simple organisms, such as bacteria, some patterns such as 'feed-forward loops' come up again and again.[7] It might be that, within an embryo, specific patterns of signal networks are always associated with specific types of event, whatever its molecular details. If this were true, it would provide a new layer of understanding, and also a new way of asking whether development of organizations larger than the individual organism really is similar, in essentials, to development of the organism itself. Similar approaches could compare the networks that connect cells within a developing organism with those that connect organisms in a developing ecosystem. It is perhaps possible that approaches like these may reveal interesting universal principles that apply across life at different scales.

Even our current understanding of embryonic development is such that it is beginning to be used to construct a new kind of medicine. People whose bodies are injured, by defective development, by trauma, or by infectious disease, are not always able to rebuild the missing tissue. Even where stem cells in some undamaged neighbouring part of the tissue are healthy, and in principle able to contribute to new growth, they may not be able to repopulate an environment ripped to shreds by inflammation and scarring. For just over half a century, surgeons have been able to treat such patients by grafting, either pieces of their own tissue (e.g. skin grafts for burns, one of the earliest applications of the technique) or tissues or organs from a recently deceased or, in some cases living, donor. Examples include grafts of kidney, heart, and lungs. The application of the technique is limited because the trauma of an operation causes damaged tissues to release alarm signals, activating defensive responses and causing phagocytes to collect fragments of cells from the new tissue and present them to T cells. If the T cells recognize anything that activates them, such as a protein structure not present in the recipient and therefore not used to eliminate young T cells that react to it in the thymus (Chapter 17), they will organize a response against the organ and it will be destroyed ('rejected'). People needing a transplant, therefore, have to wait until one becomes available from someone with their own tissue type. In practice, this means that most people wait for years, kept alive by inconvenient and imperfect life-support machines (such as dialysis units). It would be very helpful if we could build new tissues and organs without having to harvest them from other people.

A linear view of development, in which genes operate in fixed sequence according to a blueprint, suggests little hope of building new tissues from scratch without beginning with an embryo. Doing that would effectively mean creating a new human foetus or baby just to use it for spare parts, a concept morally repugnant and out of the question in civilized society. A view of development that takes into account the real actions of genes and their products in making mechanisms that allow molecules and cells to communicate and to organize themselves, on the other hand, gives a far more optimistic view. If cells can organize themselves into tissues, making correct 'decisions' based on what they find around them, in normal embryonic development, can we persuade them to perform the same tricks in the artificial environment of a culture dish?

The answer seems to be yes. One piece of evidence comes from work on the kidney, development of which was described in Chapter 10. It is possible, with the aid of digestive enzymes, to disrupt the delicate architecture of developing kidney tissue so completely that it is reduced to a cloud of individual cells floating about freely in a test tube. If these cells are then brought back together into a random mass, they move about spontaneously, each locating others of its type. Over the course of a few days they organize themselves, with no external help from experimenters, into something that is essentially indistinguishable from normal developing kidney tissue.[8, 9] When this is viewed first hand down a microscope as it actually happens, it speaks more eloquently than words ever can of our cells' innate ability to communicate and organize themselves, even in a bizarre and artificial situation.

It is a long way from these early laboratory explorations with kidneys, lungs, and the various other tissues to which the technique has now been applied, by my own laboratory and by others, to making practical organs as complex as the kidney that can be transplanted into a human in need: probably decades. Nevertheless, the growing appreciation that cells use signals and feedback to organize themselves into appropriate structures in the embryo, and that they can do the same thing even in the test tube if they are handled properly, provides a promising avenue for research. Stem cells, in particular, being able to produce all of the required mature cells of a tissue, may offer a very powerful means to generate new tissues if we can set them up so that their self-organizing mechanisms are activated. The use of bone-marrow-derived stem cells to replace the blood and immune systems of someone whose own has been destroyed by anti-leukaemia therapy has already become relatively routine: the simple structure of bone marrow lends itself to this. Stem cells resident in the skin, which normally replace skin and hair cells, have now been used successfully to treat massive burns that have destroyed a patient's own skin stem cells. Mesenchymal stem cells from a patient's own bone marrow have also been used to repopulate donated connective tissue, washed free of a donor's cells, with a patient's own cells. This effectively makes the piece of connective tissue appear to the patient as his own when it is grafted inside him, so it will not be rejected. A celebrated early example of this type of operation[10] was the reconstruction of 'Claudia's trachea', the windpipe of a woman who had lost her own to disease (the case was widely covered in newspapers at the time). There is still much to be learned and much to do, and the gap between the real state of knowledge and the hype that appears in many popular newspapers remains large, but we are probably now living through a medical revolution, founded on our improved understanding of how we come to be. Progress is not guaranteed: it depends critically on continued support for research by hard-pressed taxpayers, and a continuing supply of enthusiastic young scientists wanting to devote their lives to discovering humanity's hidden secrets.

Even if an improved understanding of development yields a greatly improved ability to repair the effects of injury and disease, we must face the fact that our body maintenance systems are not perfect. Stem cells do a great job, but slowly errors creep into the system and toxins accumulate, sowing confusion and blocking normal lines of communication and cells' abilities to respond correctly. The loss of performance is not noticeable at first, usually not for many decades, but as trivial instances of damage accumulate, they impact on the ability of the body to conduct proper repairs. This means less efficient physiology, which means even less efficient repair. This is positive feedback, and in this context it works against us: it makes us, however careful we are, mortal. Our genes, the genes that specified production of the protein machines that built us, can certainly be passed on and, with genes from another individual, they will begin again the task of building a fresh body—another, younger mortal. We tell ourselves that this is a 'life cycle', but it is not. From the point of view of genes, life may be cyclical but, from the point of view of individual humans, it is a one-way journey: the 'cycle of life' is just a fiction invented by naked apes to comfort themselves against fear of the dark.

We know we pass this way but once. Understanding more about how we got here, about the astonishing processes by which we constructed ourselves, only adds to the respect and awe with which we can view the creation of each human being, whether a stranger, a friend, or our own, unique, self-made self.

GLOSSARY

In this glossary, I have tried to strike a sensible compromise between clarity and pedantry. The definitions given are adequate for the context of this book but are not necessarily complete or surrounded by the qualifications that a formal definition would require.

Actin Actin is a cellular protein that can polymerize to form microfilaments, which are important components of the cytoskeleton. A complex between actin and myosin can generate mechanical tension.

Adaptive self—organization A process by which a multitude of 'dumb' objects, following simple rules, organize themselves into a larger-scale system that has features and behaviour absent in any of its components. Adaptive self-organization is also known as 'swarm intelligence'. The process by which the high-level features arise is known as 'emergence'.

Amino-acid Amino-acids are a family of twenty small molecules that are the subunits from which proteins are made. Each of the twenty amino-acids has a similar molecular 'backbone' but carries its own unique side-chain: the sequence in which the amino-acids occur in a protein determines the shape of the protein and its interactions with other molecules. This sequence of amino acids is specified by the sequence of bases in the gene encoding the protein.

Amniotic cavity The amniotic cavity is a closed, fluid-filled sac that forms above the epiblast and continues to surround the foetus until birth.

Anencephaly Anencephaly is an abnormality that results from a failure of neural tube closure in the head region. Large parts of the brain are missing, and the inside of what brain does manage to form is open and visible at the unclosed back of the head.

Antibody A secreted protein, one end of which recognizes a specific structure such as part of a bacterium and the other end of which can activate elements of the immune system, including complement (q.v.) and phagocytes (q.v.). Each antibody has its own specific target structure.

Aorta The aorta (plur: aortae) is the main vessel that carries blood from the heart towards lesser arteries and thence the tissues. The two aortae of embryonic life undergo considerable remodelling, as described in this book, to give rise to one main aorta in the adult.

Artery A vessel carrying high-pressure blood in the direction from heart to tissues. The blood will return along veins.

Axis Axes (sing: axis) are the directions that define the body. One axis runs from head-to-tail, one from back-to-belly and one from left to right. If 'axis' is used without qualification, it normally means the head-to-tail one.

Axon The long process that conducts electrical signals from the cell body of a neuron to connect with another neuron to a muscle.

Bacterium Bacteria (sing. bacterium) are simple cells, only around 1/1000th of the volume of a typical human cell and having some unique features, such as cell walls, not found in human cells. Bacteria can be helpful companions or dangerous pathogens, depending on what they are and where in the body they are.

Base Bases are the unique part of the nucleotide subunits from which DNA and RNA are made. There are 4 bases: A, C, G, and T in the case of DNA, and A, C, G, and U in the case of RNA.

BCR BCR (B cell receptor) is NOT a receptor for B cells, but rather a molecule related closely to an antibody (q.v.), but which is part of a B cell's outer membrane. The BCR recognizes a specific structure, such as part of a bacterium. Each BCR has its own specific target structure.

BMP BMP (Bone Morphogenetic Protein) is a signalling molecule. As is the case for many other signalling molecules, this protein was named after its first-discovered effect, and it is now known to control a far larger range of developmental events than its name implies.

Brownian Motion The random motion of small objects (for example, large proteins) caused by their being jostled by molecules of the water.

Capillary Capillaries are the narrowest of the blood vessels. Of all the vessels, capillaries have the most intimate relationship with the tissues, giving them oxygen and nutrients and removing waste products.

Centrosome The centrosome is the main microtubule-organizing centre of a cell, and is located automatically in the physical centre of that cell by the mechanisms discussed in Chapter 2.

Chiasm A chiasm is a 'crossroads' within the nervous system.

Chromosome Chromosomes consist of long molecules of DNA, carrying thousands of genes, wrapped up with proteins that give the chromosome its structure. Humans have forty-six chromosomes in each of their somatic (q.v.) cells.

Cilium Cilia (sing: cilium) are short projections from the surfaces of cells. They can beat to move fluid across the cell surface, and can also be involved in cell signalling.

Cloaca The cloaca, present in the foetus but not after birth, is a common exit for alimentary, urinary, and reproductive systems. Before birth, the cloaca is divided into rectum, urethra, and (in girls) vagina.

Colliculus—see superior colliculus

Complement The complement system is a set of proteins that recognize the surfaces of common bacteria. They punch holes in these surfaces and also initiate a response by cells of the immune system.

DNA DNA (Deoxyribonucleic acid) is the polymer of nucleotides (q.v.) and is the physical substance of which genes are made.

Defensin Defensins are antibacterial proteins.

E-cadherin E-cadherin is a cell–cell adhesion molecule.

Ectoderm Ectoserm is the outer layer of the post-gastrulation embryo. Most of it will go on to form the outer layer of the skin: other parts of the ectoderm form the neural tube and some other structures.

Embryonic stem cell Embryonic stem (ES) cells are cells derived from early embryos. They can be grown in culture and are capable of producing any cell type in the body. Manipulation of ES cells is the basis of generating genetically manipulated mice.

Endoderm The endoderm is the innermost of the three germ layers: It forms the lining of the gut and of the organs that develop as side-branches of the gut.

EPH EPH proteins are receptors for Ephrin signalling proteins (q.v.).

Ephrin Ephrin proteins are signalling molecules that are detected by EPH proteins.

Epiblast The upper layer of the two-layered disc that forms the embryo shortly before gastrulation. The lower layer is the hypoblast.

ES cell—see Embryonic stem cell.

Fibronectin Fibronectin is a protein secreted by cells, and is an important component of the connective substance that surrounds them.

Floor plate The ventral-most part of the neural tube.

Filopodium Filopodia (sing: filopodium) are long, fine processes that project from the leading edge of migrating cells and neuronal growth cones. Cells use them to explore the environment for the purposes of navigation.

FGF Fibroblast growth factor: this is a signalling molecule. As is the case for many other signalling molecules, this protein was named after its first-discovered effect, and it is now known to control a far larger range of developmental events than its name implies.

Fucose A type of sugar that (like many other sugars) can be added to proteins.

Gastrulation The process by which cells of the epiblast 'dive down' to make the endoderm and mesoderm and cells that do not dive down become the ectoderm.

Gene A gene is a section of a chromosome that specifies the production of a particular molecule of RNA: in most cases, this will be mRNA (q.v.) that will in turn specify the production of a particular protein. In some cases, one gene can be used to produce more than one protein because mRNA made from the gene can be edited in different ways to yield different final mRNA 'scripts'. These different types of editing are controlled by cellular proteins, in turn produced from their own genes.

GDNF GDNF (glial cell derived neurotrophic factor) is a cell-to-cell signalling molecule. As is the case for many other signalling molecules, this protein was named after its first-discovered effect, and it is now known to control a far larger range of developmental events than its name implies.

Germ line The germ line is the sequence of cells that give rise, ultimately, to sperm and egg. The germ line is set aside early in embryogenesis and they make nothing else.

Glomerulus Glomeruli (sing: glomerulus) are the blood-filtering units of the kidney.

Gradient A gradient is a steady change in the concentration of a molecule in space, for example because it is made in one place and is spreading out from there.

Hif1α Hif1α (Hypoxia-inducible factor 1α) is a protein found inside cells. It is normally very short-lived but it survives for much longer in the absence of oxygen. This gives it the chance to activate the expression of specific genes, notably those connected with signalling for new growth of blood vessels, the arrival of which will solve the problem of there being too little oxygen.

Hormone A hormone is a signalling molecule that can work at long range, typically spreading throughout the body in the bloodstream.

HOX code The 'HOX code' is not really a code: it is a phrase used to summarize the relationship between the position of cells along the head–tail axis and the combination of HOX genes they express. Different combinations of HOX codes give rise to different cell behaviour, and thus to the formation of different anatomical objects at different levels of the head–tail axis (for example, rib-free vertebrae of the neck and rib-bearing vertebrae of the chest).

HSC HSC (haematopoietic stem cells) are the stem cells that give rise to the various cells of the blood.

Hypoblast The lower layer of the two-layered disc that forms the embryo shortly before gastrulation. The upper layer is the epiblast.

iPS cell Induced pluripotent stem cell: a normal adult cell, for example from the connective tissue of the skin, that has been treated so that it has become very similar to an embryonic stem (ES) cell, capable of making any cell type of the body.

IGF IGF (insulin-like growth factor) I and II are secreted cell–cell signalling proteins: they are important in control of growth. They are 'insulin like' in structure but do not directly control blood sugar and energy flows in the way that insulin itself does.

Integrin Integrins are a family of proteins by which cells can bind to the connective substance that surrounds them. Typically, they connect to protein complexes inside the cells as well, making a mechanical connection from the external proteins to the internal cytoskeleton. They can also initiate internal cellular signals.

Jagged Jagged is a protein that forms part of the surface of one cell and signals to a specific receptor, Notch, on a neighbouring cell.

L1CAM A cell–cell adhesion molecule.

Laminin Laminin is a protein that is secreted by cells and which is an important component of the extracellular matrix, particularly the 'basement membrane' that separates epithelia from their underlying connective tissue.

Limb bud A limb bud is an outgrowth from the body wall that will give rise to a limb.

Limbus The limbus is the border between the cornea and the sclera (the outer surface of the white of the eye).

Mesoderm The mesoderm is one of the three germ layers of the body, formed at gastrulation. The mesoderm lies between the ectoderm (q.v.) and the endoderm (q.v.).

mRNA mRNA (messenger RNA) is an RNA 'copy' of the information in a gene. mRNA is exported from the cell nucleus and used as a template for protein synthesis out in the cytoplasm.

Microfilament Microfilaments are polymers of actin: they form the main tension-bearing (and, when combined with myosin, tension-generating) element of the cytoskeleton. Sometimes, for example at the leading edge of a migrating cell, short microfilaments are used in compression.

Microtubule Microtubules are polymers of tubulin: they form the main compression-bearing elements of the cytoskeleton and also act as 'tram-tracks' along which cellular components can be moved from place to place within the cell.

Microorganism A small organism such as a bacterium, unicellular fungus, or parasite.

Motor neuron A motor neuron is a neuron that sends electrical signals to muscle cells, and is therefore responsible for driving movement (both voluntary, as in an arm, and involuntary, as in gut peristalsis).

Mucus Mucis is a slippery secretion consisting mainly of water and complex, long carbohydrate molecules. Mucus often also contains antibacterial molecules. It both washes clean and defends vulnerable surfaces of the body such as the nasal passages and vagina.

Mullerian duct Mullerian ducts are a pair of tubes, one running just to the left and one to the right of the mid-line of an early foetus, that give rise to the upper reproductive tract of females. In males, they disappear.

Mutant Classically, a mutant is any variant of an organism that has an abnormal structure, chemistry, or behaviour, whether this be because of altered genes or environmental influence. In modern use, 'mutant' is used to refer only to organisms different because they carry one or more altered genes.

Myosin Myosin is a motor protein that can interact with actin microfilaments to generate mechanical tension.

N-cadherin N-cadherin is a cell–cell adhesion molecule.

Neural crest The neural crest is a population of cells that arises in the dorsal-most part of the neural tube, leaves that tube and migrates to settle in, and make a variety of neural and non-neural tissues in, the rest of the body.

Neural plate The neural plate is a strip of ectoderm along the mid-line of the body: it invaginates to form the neural tube (q.v.).

Neural tube The neural tube is a tube running the length of the body just under the dorsal midline. It gives rise to the spinal cord and brain and also to the neural crest (q.v.).

Neurocristopathy Neurocristopathies are diseases caused by faulty development or behaviour of the neural crest (q.v.).

Neurotransmitter A neurotransmitter is a molecule that mediates signalling in a chemical synapse (q.v.).

Neurite A neurite is a long process emerging from the cell body of a neuron, carrying electrical signals either to or from that neuron.

Neuron A neuron is a cell of the nervous system that receives, processes, and transmits electrical information: it is the cell type mainly responsible for thought and action.

Node The node is a structure at the anterior end of the primitive streak The 'diving through' movements of gastrulation (q.v.) begin here.

Notch Notch is a receptor protein for cell surface-bound signalling protiens such as Jagged (q.v.). Because Jagged is part of a cell surface, Jagged-to-Notch signalling takes place between neighbouring cells.

Notochord The notochord is a thin rod running just ventral to the neural tube. It is a very important source of signals that organize the embryo, although it contributes in only a minor way to adult anatomy.

Nucleotide A nucleotide is a combination of a base (q.v.) and a sugar. The sugar structure is slightly different in DNA and RNA (which is what makes these molecules different).

Nucleus The nucleus is a membrane-bound compartment inside human cells, in which the chromosomes (q.v.) reside.

Paneth cell Paneth cells lie near the base of intestinal crypts. They are involved in protecting neighbouring cells, including stem cells, from bacterial attack.

Peer replacement Peer replacement is a method of cell renewal during tissue maintenance, in which a damaged or dead cell is replaced by multiplication of one of its similar neighbours.

Phagocyte A cell of the immune system that is specialized to engulf and destroy invaders and/or cellular debris.

Phallus In embryology, the word 'phallus' is used to refer to the common progenitor of penis and clitoris, before the stage that these structures look different in males and females. (In more general language, 'phallus' is a synonym for penis, particularly when erect, and has a variety of complex metaphorical meanings in psychoanalysis and related fields.)

Primitive streak The primitive streak is the first visible sign of the future axis of the embryo, and appears as a notch on the epiblast along the mid-line of what will be the trunk.

RA RA (retinoic acid) is a close relative of Vitamin A (from which it is made). It is a small molecule that can be used as a cell-to-cell signalling molecule.

ROBO ROBO (from 'roundabout') is the receptor for the cell signalling molecule, SLIT. One function of ROBO, in both insects and mammals, is to allow the growth cones of elongating neurites to recognize the SLIT-expressing midline of the nervous system: with no functioning ROBO, the growth cones go 'round and round' crossing and re-crossing the mid-line, hence the name.

RNA RNA (ribonucleic acid) is a long polymer of bases, related to DNA. There are various types: the one mentioned most in this book is mRNA (q.v.).

Segment Segments are repeated units along the axis of the body, which appear as variations on a common theme. Vertebrae may be considered to be segments of the trunk skeleton, for example.

SHH SHH (Sonic Hedgehog) is a secreted cell–cell signalling molecule. SLIT is a cell–cell signalling molecule detected by the receptor ROBO (q.v.).

Soma, somatic 'Soma' means 'body', and 'somatic' means 'of the body'. In an embryological context, 'somatic' means 'of the body rather than of the germ line', the germ line (qv) being the only part of the embryo that can give rise to the following generation.

Somite Somites are blocks of mesoderm either side of the neural tube. They are the progenitors of vertebrae (in a slightly complicated way), and also give rise to other bones, muscles, and connective tissue of the trunk and limbs.

SOX9 SOX9 is a transcription factor, connected with sex determination (and other things).

Spina bifida Spina bifida is an abnormality in which the neural tube fails to close, leaving the inside of the spinal cord open.

SRY SRY (sex-determining region of the Y chromosome) is the gene that directs an embryo to choose male rather than female development.

Superior colliculus The superior colliculus is a part of the brain to which nerves carrying visual information from the eyes connect. The relative positions of axon terminations in the superior colliculus reflects the relative spacing of the origins of these axons in the retina: the electrical image of the world the eyes see is therefore reproduced on the superior colliculus.

Survival factor Survival factors are signalling molecules (in the broadest sense of that phrase), secreted from one cell type, that dissuade another cell type from undergoing elective cell death ('cell suicide').

Synapse A synapse is a connection from a neurite (q.v.) of one neuron (q.v.) to another neuron or muscle. There are two types of synapse: chemical, in which signals are passed by the transmitting cell releasing some neurotransmitter to the receiving cell, and electrical, in which electrical signals are passed directly.

TCRT Cell receptors (TCR) are not receptors for T cells, but rather receptors carried by T cells that recognize a specific structure such as part of a bacterium. Each TCR has its own specific target structure.

Trophoblast Trophoblast cells form the outer layer of the blastocyst, and will go on to form embryonic components of the placenta.

Trophectoderm Trophectoderm is the name given to the trophoblast layer (q.v.) after the embryo proper has undergone gastrulation.

Trophic hypothesis The trophic hypothesis proposes that, beyond the earliest stages of embryogenesis, cells depend for their survival on signals (survival factors) made by other types of cell. This is one way in which the populations of different cells remain in proportion to one another and any cells that end up in the wrong place are eliminated.

Tubulin Tubulin is the protein from which microtubules are made.

Protein Proteins are polymers composed of amino-acids. They are the main structural components of the cell and also catalyze most of its biochemical reactions.

Vein Veins are vessels that carry blood from capillaries in the tissues back towards the heart.

VEGF VEGF (vascular endothelial growth factor) is a cell–cell signalling molecule that is a powerful trigger for blood vessel growth.

Virus A virus is a parasitic microorganism consisting of genetic material contained in a protein coat or an enveloping membrane: a virus is too simple to run metabolism of its own but, if it can arrange for its genetic material to be taken up by a human cell, this material will take control of the cell and force it to make more copies of the virus. Some viruses can remain quiescent in their host cells, lying low for years (chicken pox is an example: the virus causes shingles when it re-awakens).

Vitruvian In the context of this book, a 'Vitruvian' body is one that, though possibly unusually small or large, has the normal proportions of body parts as described by Vitruvius (and drawn in Da Vinci's sketch 'Vitruvian Man').

Wolffian duct Wolffian ducts are a pair of tubes that run, one each side of the midline of the body, from the primitive kidneys to the cloaca. They disappear almost completely from girls but are retained to form elements of the reproductive system in boys.

WNT The name WNT is a hybrid of earlier gene names and is not an acronym. WNT proteins are secreted cell–cell signalling molecules.

WT1 WT1 (Wilms tumor 1) is a protein that controls expression of specific genes in a variety of ways (by determining whether the genes are read and also by altering the way that their mRNA is edited). Loss of WT1 is associated with the childhood kidney cancer, Wilms tumour, hence the name. WT1 is also involved in sex determination.

TECHNICAL REFERENCES

Chapter 2: From One Cell to Many

1. Inoué S, Salmon ED. Force generation by microtubule assembly/disassembly in mitosis and related movements. Mol Biol Cell. 1995;6:1619–40.
2. Schatten H. The mammalian centrosome and its functional significance. Histochem Cell Biol. 2008;192:667–86.
3. Reinsch S, Gönczy P. Mechanisms of nuclear positioning. J Cell Sci. 1998;111:2283–95.
4. Holy TE, Dogterom M, Yurke B, Leibler S. Assembly and positioning of microtubule asters in microfabricated chambers. Proc. Natl. Acad. Sci. USA 1997;94:6228–31.
5. Grill SW, Hyman AA. Spindle positioning by cortical pulling forces. Dev Cell. 2005;8:461–5.
6. Kimura A, Onami S. Local cortical pulling-force repression switches centrosomal centration and posterior displacement in C. elegans. J Cell Biol. 2007;178:1347–54.
7. Kimura A, Onami S. Computer simulations and image processing reveal length-dependent pulling force as the primary mechanism for C. elegans pronuclear migration. Dev Cell. 2005;8:765–75.
8. Vallee RB, Stehman SA. How dynein helps the cell find its center: A servomechanical model. Trends Cell Biol. 2005;15:288–94.
9. Grill SW, Howard J, Schäffer E, Stelzer EH, Hyman AA. The distribution of active force generators controls mitotic spindle position. Science. 2003;301:518–21.
10. Bornens M. Centrosome composition and microtubule anchoring mechanisms. Curr Opion Cell Biol. 2002;14:25–34.
11. Yasuda S, Oceguera-Yanez F, Kato T, Okamoto M, Yonemura S, Terada Y, Ishizaki T, Narumiya S. Cdc42 and mDia3 regulate microtubule attachment to kinetochores. Nature. 2004;428:767–71.
12. Li X, Nicklas RB. Mitotic forces control a cell-cycle checkpoint. Nature. 1995; 373: 630–2.
13. Lampson MA, Renduchitala K, Khodjakov A, Kapoor TM. Correcting improper chromosome-spindle attachments during cell division. Nat Cell Biol. 2004;6:232–7.
14. Waters JC, Cole RW, Rieder CL. The force-producing mechanism for centrosome separation during spindle formation in vertebrates is intrinsic to each aster. J Cell Biol. 1993;122:361–72.

Chapter 3: Making a Difference

1. Braude P, Bolton V, Moore S. Human gene expression first occurs between the four- and eight-cell stages of preimplantation development. Nature. 1988;332:459–61.

2. Van de Velde H, Cauffman G, Tournaye H, Devroey P, Liebaers I. The four blastomeres of a 4-cell stage human embryo are able to develop individually into blastocysts with inner cell mass and trophectoderm. Hum Reprod 2008;23:1742–7.

3. Sasaki H. Mechanisms of trophectoderm fate specification in preimplantation mouse development. Dev Growth Differ. 2010;52:263–73.

4. Cohen M, Meisser A, Bischof P. Metalloproteinases and human placental invasiveness. Placenta. 2006;27:783–93.

5. Mor G. Inflammation and pregnancy: the role of toll-like receptors in trophoblast-immune interaction. Ann N Y Acad Sci. 2008;1127:121–8.

6. Shaw JL, Dey SK, Critchley HO, Horne AW. Current knowledge of the aetiology of human tubal ectopic pregnancy. Hum Reprod Update. 2010 July–August; 16(4): 432–44.

7. Maximow A A. The lymphocyte is a stem cell, common to different blood elements in embryonic development and during the post-fetal life of mammals. Eng. Trans in Cell Ther Transplant. 2009;1:e.000032.01. doi:10.3205/ctt-2009-en-000032.01.

8. Evans MJ, Kaufman MH. Establishment in culture of pluripotential cells from mouse embryos. Nature. 1981;292:154–6.

9. Pitera JE, Turmaine M, Woolf AS, Scambler PJ. Generation of mice with a conditional null fraser syndrome 1 (Fras1) allele. Genesis 2012 June 22. doi: 10.1002/dvg.22045.

10. Thomson JA, Odorico JS. Human embryonic stem cell and embryonic germ cell lines. Trends Biotechnol. 2000;18:53–7.

11. Takahashi K, Yamanaka S. Induction of pluripotent stem cells from mouse embryonic and adult fibroblast cultures by defined factors. Cell. 2006 Aug 25;126(4):663–76.

12. Cockburn K, Rossant J. Making the blastocyst: lessons from the mouse. J Clin Invest. 2010;120:995–1003.

13. Gardner RL, Rossant J. Investigation of the fate of 4–5 day post-coitum mouse inner cell mass cells by blastocyst injection. J Embryol Exp Morphol. 1979;52:141–52.

14. Lawson KA, Meneses JJ, Pedersen RA. Clonal analysis of epiblast fate during germ layer formation in the mouse embryo. Development. 1991;113:891–911.

Chapter 4: Laying Down a Body Plan

1. Thomas PQ, Brown A, Beddington RS. Hex: A homeobox gene revealing peri-implantation asymmetry in the mouse embryo and an early transient marker of endothelial cell precursors. Development. 1998; 125:85–94.

2. Bouwmeester T, Kim S, Sasai Y, Lu B, De Robertis EM. Cerberus is a head-inducing secreted factor expressed in the anterior endoderm of Spemann's organizer. Nature 1996;382:595–601.

3. Srinivas S, Rodriguez T, Clements M, Smith JC, Beddington RS. Active cell migration drives the unilateral movements of the anterior visceral endoderm. Development. 2004;131:1157–64.

4. Jones CM, Broadbent J, Thomas PQ, Smith JC, Beddington RS. An anterior signalling centre in Xenopus revealed by the homeobox gene XHex. Curr Biol. 1999 Sep 9;9(17):946–54.

5. Migeotte I, Omelchenko T, Hall A, Anderson KV. Rac1-dependent collective cell migration is required for specification of the anterior-posterior body axis of the mouse. PLoS Biol. 2010 Aug 3;8(8):e1000442.

6. Beddington RS, Robertson EJ. Axis development and early asymmetry in mammals. Cell. 1999;96:195–209.

7. Idkowiak J, Weisheit G, Plitzner J, Viebahn C. Hypoblast controls mesoderm generation and axial patterning in the gastrulating rabbit embryo. Dev Genes Evol. 2004; 214: 591–605.

8. Martinez-Barbera JP, Beddington RS. Getting your head around Hex and Hesx1: forebrain formation in mouse. Int J Dev Biol. 2001;45:327–36.

9. Voiculescu O, Bertocchini F, Wolpert L, Keller RE, Stern CD. The amniote primitive streak is defined by epithelial cell intercalation before gastrulation. Nature. 2007 Oct 25;449(7165):1049–52.

10. Azar Y, Eyal-Giladi H. Interaction of epiblast and hypoblast in the formation of the primitive streak and the embryonic axis in chick, as revealed by hypoblast-rotation experiments. J Embryol Exp Morphol. 1981;61:133–44.

11. Martin HE. Chang and Eng Bunker, "The original Siamese twins": Living, dying, and continuing under the spectator's gaze. J Am Cult. 2011;34(4):372–90.

12. Chichester P. Eng and Chang Bunker: A hyphenated life. Blur Right Country magazine 2009;17 Feb: http://blueridgecountry.com/archive/a-hyphenated-life.html.

13. Buffetaut E, Li J, Tong H, Zhang H. A two-headed reptile from the Cretaceous of China. Biol Lett. 2007;3: 80–1.

14. Oki S, Kitajima K, Meno C. Dissecting the role of Fgf signaling during gastrulation and left-right axis formation in mouse embryos using chemical inhibitors. Dev Dyn. 2010;239:1768–78.

15. Weng W, Stemple DL. Nodal signaling and vertebrate germ layer formation. Birth Defects Res C Embryo Today. 2003;69:325–32.

16. Vincent SD, Dunn NR, Hayashi S, Norris DP, Robertson EJ. Cell fate decisions within the mouse organizer are governed by graded Nodal signals. Genes Dev. 2003;17:1646–62.

17. Tam PP, Behringer RR. Mouse gastrulation: The formation of a mammalian body plan. Mech Dev. 1997;68:3–25.

18. Rossant J, Tam PP. Blastocyst lineage formation, early embryonic asymmetries and axis patterning in the mouse. Development. 2009;136:701–13.

19. Wittler L, Kessel M. The acquisition of neural fate in the chick. Mech Dev. 2004;121:1031–42.

20. Chapman SC, Matsumoto K, Cai Q, Schoenwolf GC. Specification of germ layer identity in the chick gastrula. BMC Dev Biol. 2007;7:91.

21. Gerhart J, Neely C, Elder J, Pfautz J, Perlman J, Narciso L, Linask KK, Knudsen K, George-Weinstein M. Cells that express MyoD mRNA in the epiblast are stably committed to the skeletal muscle lineage. J Cell Biol. 2007 Aug 13;178(4):649–60.

22. Streit A, Berliner AJ, Papanayotou C, Sirulnik A, Stern CD. Initiation of neural induction by FGF signalling before gastrulation. Nature. 2000;406:74–8.

23. Sausedo RA, Schoenwolf GC. Quantitative analyses of cell behaviors underlying notochord formation and extension in mouse embryos. Anat Rec. 1994;239:103–12.

24. Sulik K, Dehart DB, Iangaki T, Carson JL, Vrablic T, Gesteland K, Schoenwolf GC. Morphogenesis of the murine node and notochordal plate. Dev Dyn. 1994 Nov;201(3):260–78.

25. Jurand A. Some aspects of the development of the notochord in mouse embryos. J Embryol Exp Morphol. 1974;32:1–33.

26. McCann MR, Tamplin OJ, Rossant J, Séguin CA. Tracing notochord-derived cells using a Noto-cre mouse: implications for intervertebral disc development. Dis Model Mech. 2012;5:73–82.
27. Lee JD, Anderson KV. Morphogenesis of the node and notochord: The cellular basis for the establishment and maintenance of left–right asymmetry in the mouse. Dev Dyn. 2008;237:3464–76.
28. Santos N, Reiter JF. Tilting at nodal windmills: Planar cell polarity positions cilia to tell left from right. Dev Cell. 2010;19:5–6.
29. Hirokawa N, Tanaka Y, Okada Y, Takeda S. Nodal flow and the generation of left–right asymmetry. Cell. 2006;125:33–45.
30. Shields AR, Fiser BL, Evans BA, Falvo MR, Washburn S, Superfine R. Biomimetic cilia arrays generate simultaneous pumping and mixing regimes. Proc Natl Acad Sci U S A. 2010;107:15670–5.

Chapter 5: Beginning a Brain

1. Bertet C, Sulak L, Lecuit T. Myosin-dependent junction remodelling controls planar cell intercalation and axis elongation. Nature. 2004;429:667–71.
2. Rauzi M, Lenne PF, Lecuit T. Planar polarized actomyosin contractile flows control epithelial junction remodelling. Nature. 2010;468:1110–14.
3. Wang J, Hamblet NS, Mark S, Dickinson ME, Brinkman BC, Segil N, Fraser SE, Chen P, Wallingford JB, Wynshaw-Boris A. Dishevelled genes mediate a conserved mammalian PCP pathway to regulate convergent extension during neurulation. Development. 2006;133:767–78.
4. Lee CC, Liu KL, Tsang YM, Chen SJ, Liu HM. Fetus in fetu in an adult: diagnosis by computed tomography imaging. J Formos Med Assoc. 2005;104:203–5.
5. Kinoshita N, Sasai N, Misaki K, Yonemura S. Apical accumulation of Rho in the neural plate is important for neural plate cell shape change and neural tube formation. Mol Biol Cell. 2008;19:2289–99.
6. Saucedo RA, Smith JL, Schoenwolf GC Role of nonrandomly oriented cell division in shaping and bending of the neural plate, J. Comp Neurol 1997;381:473–88.
7. Hibbard BM. The role of folic acid in pregnancy, with particular reference to aneamia, abruption and abortion. J Obstet Gynaecol Br Commonw. 1964;71:529–42.
8. Pitkin RM. Folate and neural tube defects Am. J. Clin. Nutr. 2007;85:285S–8S.
9. Kibar Z, Capra V, Gros P. Toward understanding the genetic basis of neural tube defects. Clin. Genet. 2007;71:295–310.
10. Sano K. Intracranial dysembryogenetic tumors: Pathogenesis and their order of malignancy. Neurosurg Rev. 2001;24:162–7.
11. Afshar F, King TT, Berry CL. Intraventricular fetus-in-fetu. J Neurosurg. 1982;56:845–9.
12. Lee C, Scherr HM, Wallingford JB. Shroom family proteins regulate gamma-tubulin distribution and microtubule architecture during epithelial cell shape change. 2007;134:1431–41.

Chapter 6: Long Division

1. Glazier J A, Zhang Y, Swat M, Zaitlen B, Schnell S. Coordinated action of N-CAM, N-cadherin, EphA4, and ephrinB2 translates genetic prepatterns into structure during somitogenesis in chick. Curr Top Dev Biol. 2008;81:205–47.

2. Dubrulle J, McGrew MJ, Pourquié O. FGF signaling controls somite boundary position and regulates segmentation clock control of spatiotemporal Hox gene activation. Cell. 2001;106:219–32.

3. Naiche LA, Holder N, Lewandoski M. FGF4 and FGF8 comprise the wavefront activity that controls somitogenesis. Proc Natl Acad Sci U S A. 2011;108:4018–23.

4. Aulehla A, Pourquié O. Signaling gradients during paraxial mesoderm development. Cold Spring Harb Perspect Biol. 2010;2:a000869.5.

5. J. Cooke, E.C. Zeeman. A clock and wavefront model for control of the number of repeated structures during animal morphogenesis. J Theor Biol, 1976;58: 455–76.

6. Saga Y. The mechanism of somite formation in mice. Curr Opin Genet Dev. 2012 June 26. [Epub ahead of print]

7. Gomez C, Ozbudak EM, Wunderlich J, Baumann D, Lewis J, Pourquié O. Control of segment number in vertebrate embryos. Nature. 2008;454:335–9.

8. Lynch VJ, Roth JJ, Wagner GP. Adaptive evolution of Hox-gene homeodomains after cluster duplications. BMC Evol Biol. 2006;6:86.

9. Chambeyron S, Bickmore WA. Chromatin decondensation and nuclear reorganization of the HoxB locus upon induction of transcription. Genes Dev. 2004; 18: 1119–30.

10. Sessa L, Breiling A, Lavorgna G, Silvestri L, Casari G, Orlando V. Noncoding RNA synthesis and loss of Polycomb group repression accompanies the colinear activation of the human HOXA cluster. RNA. 2007;13:223–39.

11. Chambeyron S, Da Silva NR, Lawson KA, Bickmore WA. Nuclear re-organisation of the Hoxb complex during mouse embryonic development. Development. 2005; 132: 2215–23.

12. Wellik DM. Hox patterning of the vertebrate axial skeleton. Dev Dyn. 2.

Chapter 7: Fateful Conversations

1. Brown M, Keynes R, Lumsden A. (2000) The developing brain. Oxford University Press.

2. Ulloa F, Briscoe J. (2007) Morphogens and the control of cell proliferation and patterning in the spinal cord. Cell Cycle. 2007 November 1;6(21): 2640–9.

3. Goulding MD, Lumsden A, Gruss P. (1993) Signals from the notochord and floor plate regulate the region-specific expression of two Pax genes in the developing spinal cord. Development. 1993 117: 1001–16.

4. Yamada T, Pfaff SL, Edlund T, Jessell TM. (1993) Control of cell pattern in the neural tube: Motor neuron induction by diffusible factors from notochord and floor plate. Cell. 1993 May 21;73(4): 673–86.

5. Dessaud E, McMahon AP, Briscoe J. (2008) Pattern formation in the vertebrate neural tube: a sonic hedgehog morphogen-regulated transcriptional network. Development. 135: 2489–503.

6. Lee KJ, Jessell TM. The specification of dorsal cell fates in the vertebrate central nervous system. Annu Rev Neurosci. 1999; 22: 261–94.

7. Le Dréau G, Martí E. Dorsal-ventral patterning of the neural tube: A tale of three signals. Dev Neurobiol. 2012 December;72(12):1471–81.

8. Geetha-Loganathan P, Nimmagadda S, Scaal M, Huang R, Christ B. (2008) Wnt signaling in somite development. Ann Anat. 2008;190(3): 208–22.

9. Hirsinger E, Jouve C, Malapert P, Pourquié O. (1998) Role of growth factors in shaping the developing somite. Mol Cell Endocrinol. 140: 83–7.

10. Cairns DM, Sato ME, Lee PG, Lassar AB, Zeng L. A gradient of Shh establishes mutually repressing somitic cell fates induced by Nkx3.2 and Pax3. Dev Biol. 2008 November 15;323(2):152–65.

Chapter 8: Inner Journeys

1. Mullins RD, Heuser JA, Pollard TD. The interaction of Arp2/3 complex with actin: nucleation, high affinity pointed end capping, and formation of branching networks of filaments. Proc Natl Acad Sci U S A. 1998;95: 6181–6.

2. Abraham VC, Krishnamurthi V, Taylor DL, Lanni F. The actin-based nanomachine at the leading edge of migrating cells. Biophys J. 1999 September;77(3): 1721–32.

3. Maly IV, Borisy GG. Self-organization of a propulsive actin network as an evolutionary process. Proc Natl Acad Sci U S A. 2001 September 25;98(20): 11324–9.

4. Beningo KA, Dembo M, Kaverina I, Small JV, Wang YL. Nascent focal adhesions are responsible for the generation of strong propulsive forces in migrating fibroblasts. J Cell Biol. 2001;153: 881–8.

5. Miao L, Vanderlinde O, Stewart M, Roberts TM. Retraction in amoeboid cell motility powered by cytoskeletal dynamics. Science. 2003;302: 1405–7.

6. Pelham RJ Jr, Wang Y. High resolution detection of mechanical forces exerted by locomoting fibroblasts on the substrate. Mol Biol Cell. 1999;10: 935–45.

7. Suter DM, Errante LD, Belotserkovsky V, Forscher P. The Ig superfamily cell adhesion molecule, apCAM, mediates growth cone steering by substrate-cytoskeletal coupling. J Cell Biol. 1998 April 6;141(1): 227–40.

8. Poliakov A, Cotrina M, Wilkinson DG. Diverse roles of eph receptors and ephrins in the regulation of cell migration and tissue assembly. Dev Cell. 2004;7: 465–80.

9. Gammill LS, Gonzalez C, Gu C, Bronner-Fraser M. Guidance of trunk neural crest migration requires neuropilin 2/semaphorin 3F signaling. Development. 2006;133: 99–106.

10. Young HM, Anderson RB, Anderson CR. Guidance cues involved in the development of the peripheral autonomic nervous system. Auton Neurosci. 2004;112: 1–14.

11. Huber K. The sympathoadrenal cell lineage: Specification, diversification, and new perspectives. Dev Biol. 2006;298: 335–43.

12. Belmadani A, Tran PB, Ren D, Assimacopoulos S, Grove EA, Miller RJ. The chemokine stromal cell-derived factor-1 regulates the migration of sensory neuron progenitors. J Neurosci. 2005;25: 3995–4003.

13. Santiago A, Erickson CA. Ephrin-B ligands play a dual role in the control of neural crest cell migration. Development. 2002;129: 3621–32.

14. Erickson CA, Goins TL. Avian neural crest cells can migrate in the dorsolateral path only if they are specified as melanocytes. Development. 1995;121: 915–24.

15. Anderson DJ. Genes, lineages and the neural crest: A speculative review. Philos Trans R Soc Lond B Biol Sci. 2000;355: 953–64.

16. Amiel J, Sproat-Emison E, Garcia-Barcelo M, Lantieri F, Burzynski G, Borrego S, Pelet A, Arnold S, Miao X, Griseri P, Brooks AS, Antinolo G, de Pontual L, Clement-Ziza M, Munnich A, Kashuk C, West K, Wong KK, Lyonnet S, Chakravarti A, Tam PK, Ceccherini I, Hofstra RM, Fernandez R. Hirschsprung disease, associated syndromes and genetics: A review. J Med Genet. 2008;45: 1–14.

17. Iso M, Fukami M, Horikawa R, Azuma N, Kawashiro N, Ogata T. (2008) SOX10 muta-tion in Waardenburg syndrome type II. Am J Med Genet. 2008;146A: 2162–3.
18. Sznajer Y, Coldéa C, Meire F, Delpierre I, Sekhara T, Touraine RL. A *de novo* SOX10 mu-tation causing severe type 4 Waardenburg syndrome without Hirschsprung disease. Am J Med Genet. 2008;146A: 1038–41.
19. Yang SZ, Cao JY, Zhang RN, Liu LX, Liu X, Zhang X, Kang DY, Li M, Han DY, Yuan HJ, Yang WY. Nonsense mutations in the PAX3 gene cause Waardenburg syndrome type I in two Chinese patients. Chin Med J (Engl). 2007;120: 46–9.
20. Ohtani S, Shinkai Y, Horibe A, Katayama K, Tsuji T, Matsushima Y, Tachibana M, Kunieda T. A Deletion in the Endothelin-B Receptor Gene is Responsible for the Waardenburg Syndrome-Like Phenotypes of WS4 Mice. Exp Anim. 2006; 55: 491–5.
21. Dixon J, Jones NC, Sandell LL, Jayasinghe SM, Crane J, Rey JP, Dixon MJ, Trainor PA. Tcof1/Treacle is required for neural crest cell formation and proliferation deficiencies that cause craniofacial abnormalities. Proc Natl Acad Sci U S A. 2006;103: 13403–8.
22. Sakai D, Trainor PA. Treacher Collins syndrome: Unmasking the role of Tcof1/treacle. Int J Biochem Cell Biol. 2009;41: 1229–32.

Chapter 9: Plumbing

1. Lucretius, 'On The Nature of Things', translated by William Ellery Leonard.
2. Sabin FR. Studies on the origin of blood vessels and of red blood corpuscles as seen in the living blastoderm of the chick during the second day of incubation. Carnegie Contrib Embryol 1920;9: 213–62.
3. Xiong JW. Molecular and developmental biology of the hemangioblast. Dev Dyn. 2008;237: 1218–31.
4. Cleaver O, Krieg PA. VEGF mediates angioblast migration during development of the dorsal aorta in Xenopus. Development. 1998;125: 3905–14.
5. Lamont RE, Childs S. MAPping out arteries and veins. Sci STKE. 2006;2006(355):pe39.
6. Poole TJ, Finkelstein EB, Cox CM. The role of FGF and VEGF in angioblast induction and migration during vascular development. Dev Dyn. 2001;220: 1–17.
7. Brown LA, Rodaway AR, Schilling TF, Jowett T, Ingham PW, Patient RK, Sharrocks AD. Insights into early vasculogenesis revealed by expression of the ETS-domain transcription factor Fli-1 in wild-type and mutant zebrafish embryos. Mech Dev. 2000;90: 237–52.
8. Vokes SA, Yatskievych TA, Heimark RL, McMahon J, McMahon AP, Antin PB, Krieg PA. Hedgehog signaling is essential for endothelial tube formation during vasculogen-esis. Development. 2004;131: 4371–80.
9. Bressan M, Davis P, Timmer J, Herzlinger D, Mikawa T. Notochord-derived BMP ant-agonists inhibit endothelial cell generation and network formation. Dev Biol. 2009;326: 101–11.
10. Garriock RJ, Czeisler C, Ishii Y, Navetta AM, Mikawa T. An anteroposterior wave of vascular inhibitor downregulation signals aortae fusion along the embryonic midline axis. Development. 2010;137: 3697–706.
11. Williams C, Kim SH, Ni TT, Mitchell L, Ro H, Penn JS, Baldwin SH, Solnica-Krezel L, Zhong TP. Hedgehog signaling induces arterial endothelial cell formation by re-pressing venous cell fate. Dev Biol. 2010;341: 196–204.

12. Marvin MJ, Di Rocco G, Gardiner A, Bush SM, Lassar AB. Inhibition of Wnt activity induces heart formation from posterior mesoderm. Genes Dev. 2001;15: 316–27.

13. Paige SL, Osugi T, Afanasiev O, Pabon L, Reinecke H, Murry CE. Endogenous Wnt/β-Catenin Signaling Is Required for Cardiac Differentiation in Human Embryonic Stem Cells. PLoS One. 2010; 5(6): e11134.

14. Forouhar AS, Liebling M, Hickerson A, Nasiraei-Moghaddam A, Tsai HJ, Hove JR, Fraser SE, Dickinson ME, Gharib M. (2006) The embryonic vertebrate heart tube is a dynamic suction pump. Science. 2006;312: 751–3.

15. Vaughan A. Signalman's morning. 1981. John Murray.

16. Makanya AN, Hlushchuk R, Djonov VG. Intussusceptive angiogenesis and its role in vascular morphogenesis, patterning, and remodeling. Angiogenesis. 2009;12: 113–23.

17. Ribatti D. Hemangioblast does exist. Leukaemia Research 2008;32:850–4.

18. Zovein AC, Hofmann JJ, Lynch M, French WJ, Turlo KA, Yang Y, Becker MS, Zanetta L, Dejana E, Gasson JC, Tallquist MD, Iruela-Arispe ML. Fate tracing reveals the endothelial origin of hematopoietic stem cells. Cell Stem Cell. 2008;3: 625–36.

19. Peeters M, Ottersbach K, Bollerot K, Orelio C, de Bruijn M, Wijgerde M, Dzierzak E. Ventral embryonic tissues and Hedgehog proteins induce early AGM hematopoietic stem cell development. Development. 2009;136:2613–21.

20. Yoon MJ, Koo BK, Song R, Jeong HW, Shin J, Kim YW, Kong YY, Suh PG. Mind bomb-1 is essential for intraembryonic hematopoiesis in the aortic endothelium and the sub-aortic patches. Mol Cell Biol. 2008;28:4794–804.

21. André H, Pereira TS. Identification of an alternative mechanism of degradation of the hypoxia-inducible factor-1alpha. J Biol Chem. 2008;283:29375–84.

22. Qutub AA, Popel AS. Three autocrine feedback loops determine HIF1 alpha expression in chronic hypoxia. Biochim Biophys Acta. 2007;1773:1511–25.

23. Forsythe JA, Jiang BH, Iyer NV, Agani F, Leung SW, Koos RD, Semenza GL. Activation of vascular endothelial growth factor gene transcription by hypoxia-inducible factor 1. Mol Cell Biol. 1996;16:4604–13.

24. Djonov VG, Kurz H, Burri PH. Optimality in the developing vascular system: Branching remodeling by means of intussusception as an efficient adaptation mechanism. Dev Dyn. 2002;224:391–402.

Chapter 10: Organizing Organs

1. Sebinger DD, Unbekandt M, Ganeva VV, Ofenbauer A, Werner C, Davies JA. A novel, low-volume method for organ culture of embryonic kidneys that allows development of cortico-medullary anatomical organization. PLoS One. 2010 May 10;5(5):e10550.

2. Davies JA. Mechanisms of Morphogenesis. 2005; Academic Press.

3. Sainio K, Suvanto P, Davies J et al. Glial-cell-line-derived neurotrophic factor is required for bud initiation from ureteric epithelium. Development. 1997;124:4077–87.

4. Davies JA, Millar CB, Johnson EM Jr, Milbrandt J. Neurturin: An autocrine regulator of renal collecting duct development. Dev Genet. 1999;24(3–4):284–92.

5. Moore MW, Klein RD, Fariñas I, Sauer H, Armanini M, Phillips H, Reichardt LF, Ryan AM, Carver-Moore K, Rosenthal A. Renal and neuronal abnormalities in mice lacking GDNF. Nature. 1996;382(6586):76–9.

6. Michael L, Davies JA. Pattern and regulation of cell proliferation during murine ureteric bud development. J Anat. 2004;204:241–55.

7. Carroll TJ, Park JS, Hayashi S, Majumdar A, McMahon AP. Wnt9b plays a central role in the regulation of mesenchymal to epithelial transitions underlying organogenesis of the mammalian urogenital system. Dev Cell. 2005;9:283–92.

8. Nelson CM, Vanduijn MM, Inman JL, Fletcher DA, Bissell MJ. Tissue geometry determines sites of mammary branching morphogenesis in organotypic cultures. Science. 2006;314:298–300.

9. Lee WC, Davies JA. Epithelial branching: The power of self-loathing. Bioessays. 2007;29:205–7.

10. Tufro A. VEGF spatially directs angiogenesis during metanephric development in vitro. Dev Biol. 2000;227:558–66.

11. Davies JA. Inverse Correlation Between an Organ's Cancer Rate and Its Evolutionary Antiquity. Organogenesis. 2004;1:60–3.

12. Vaccari B, Mesquita FF, Gontijo JA, Boer PA. Fetal kidney programming by severe food restriction: Effects on structure, hormonal receptor expression and urinary sodium excretion in rats. J Renin Angiotensin Aldosterone Syst. 2013 March 12.

13. Dötsch J, Plank C, Amann K. Fetal programming of renal function. Pediatr Nephrol. 2012;27:513–20.

14. Gluckman PD, Hanson MA, Cooper C, Thornburg KL. Effect of in utero and early-life conditions on adult health and disease. N Engl J Med. 2008;359:61–73.

Chapter 11: Taking up Arms (and Legs)

1. King M, Arnold JS, Shanske A, Morrow BE. T-genes and limb bud development. Am J Med Genet A. 2006;140:1407–13.

2. Takeuchi JK, Koshiba-Takeuchi K, Suzuki T, Kamimura M, Ogura K, Ogura T. Tbx5 and Tbx4 trigger limb initiation through activation of the Wnt/Fgf signaling cascade. Development. 2003;130:2729–39.

3. Kawakami Y, Capdevila J, Büscher D, Itoh T, Rodríguez Esteban C, Izpisúa Belmonte JC. WNT signals control FGF-dependent limb initiation and AER induction in the chick embryo. Cell. 2001;104:891–900.

4. Cohn MJ, Izpisúa-Belmonte JC, Abud H, Heath JK, Tickle C. Fibroblast growth factors induce additional limb development from the flank of chick embryos. Cell. 1995;80:739–46.

5. Ohuchi H, Nakagawa T, Yamauchi M, Ohata T, Yoshioka H, Kuwana T, Mima T, Mikawa T, Nohno T, Noji S. An additional limb can be induced from the flank of the chick embryo by FGF4. Biochem Biophys Res Commun. 1995;209:809–16.

6. Kawakami Y, Capdevila J, Büscher D, Itoh T, Rodríguez Esteban C, Izpisúa Belmonte JC. WNT signals control FGF-dependent limb initiation and AER induction in the chick embryo. Cell. 2001;104:891–900.

7. Crossley PH, Martin GR. The mouse Fgf8 gene encodes a family of polypeptides and is expressed in regions that direct outgrowth and patterning in the developing embryo. Development. 1995;121:439–51.

8. Nikbakht N, McLachlan JC. A proximo-distal gradient of FGF-like activity in the embryonic chick limb bud. Cell Mol Life Sci. 1997;53:447–51.

9. Summerbell D, Lewis JH, Wolpert L. Positional information in chick limb morphogenesis. Nature. 1973;244:492–6.

10. Wolpert L, Tickle C, Sampford M. The effect of cell killing by x-irradiation on pattern formation in the chick limb. J Embryol Exp Morphol. 1979;50:175–93.

11. Galloway JL, Delgado I, Ros MA, Tabin CJ. A reevaluation of X-irradiation-induced phocomelia and proximodistal limb patterning. Nature. 2009;460(7253):400–4.

12. Cooper KL, Hu JK, ten Berge D, Fernandez-Teran M, Ros MA, Tabin CJ. Initiation of proximal-distal patterning in the vertebrate limb by signals and growth. Science. 2011;332:1083–6.

13. Roselló-Díez A, Ros MA, Torres M. Diffusible signals, not autonomous mechanisms, determine the main proximodistal limb subdivision. Science. 2011;332:1086–8.

14. There is recent evidence that retinoic acid itself may not be required for limb patterning (although it can drive it experimentally, as explained in the main text). This suggests the presence of some additional, unidentified signal that spreads from the flank in a similar way to retinoic acid and affects the same pathway that retinoic acid affects: see Zhao X, Sirbu IO, Mic FA, Molotkova N, Molotkov A, Kumar S, Duester G. Retinoic acid promotes limb induction through effects on body axis extension but is unnecessary for limb patterning. Curr Biol. 2009;19:1050–7.

15. Vargesson N, Kostakopoulou K, Drossopoulou G, Papageorgiou S, Tickle C. Characterisation of hoxa gene expression in the chick limb bud in response to FGF. Dev Dyn. 2001;220:87–90.

16. Abbasi AA. Evolution of vertebrate appendicular structures: Insight from genetic and palaeontological data. Dev Dyn. 2011;240:1005–16.

17. Altabef M, Tickle C. Initiation of dorso-ventral axis during chick limb development. Mech Dev. 2002;116:19–27.

18. Parr BA, McMahon AP. Dorsalizing signal Wnt-7a required for normal polarity of D-V and A-P axes of mouse limb. Nature. 1995;374:350–3.

19. Zeller R, López-Ríos J, Zuniga A. Vertebrate limb bud development: moving towards integrative analysis of organogenesis. Nat Rev Genet. 2009;10:845–58.

20. Therapontos C, Erskine L, Gardner ER, Figg WD, Vargesson N. Thalidomide induces limb defects by preventing angiogenic outgrowth during early limb formation. Proc Natl Acad Sci USA. 2009;106:8573–8.

Chapter 12: The Y and How

1. Ginsburg M, Snow MH, McLaren A. Primordial germ cells in the mouse embryo during gastrulation. Development. 1990;110:521–8.

2. Lawson KA, Hage WJ. Clonal analysis of the origin of primordial germ cells in the mouse. Ciba Found Symp. 1994;182:68–91.

3. Bradford ST, Wilhelm D, Bandiera R, Vidal V, Schedl A, Koopman P. A cell-autonomous role for WT1 in regulating Sry in vivo. Hum Mol Genet. 2009;18:3429–38.

4. Sekido R, Bar I, Narváez V, Penny G, Lovell-Badge R. SOX9 is up-regulated by the transient expression of SRY specifically in Sertoli cell precursors. Dev Biol. 2004; 274: 271–9.

5. Piprek RP. Genetic mechanisms underlying male sex determination in mammals. J Appl Genet. 2009;50:347–60.

6. Barrionuevo F, Bagheri-Fam S, Klattig J, Kist R, Taketo MM, Englert C, Scherer G. Homozygous Inactivation of Sox9 Causes Complete XY Sex Reversal in Mice. Biol Reprod 2006;74,195–201.

7. Vidal VP, Chaboissier MC, de Rooij DG, Schedl A. Sox9 induces testis development in XX transgenic mice. Nat Genet. 2001;2:216–17.

8. Kim Y, Kobayashi A, Sekido R, DiNapoli L, Brennan J, Chaboissier MC, Poulat F, Behringer RR, Lovell-Badge R, Capel B. Fgf9 and Wnt4 act as antagonistic signals to regulate mammalian sex determination. PLoS Biol. 2006;4:e187.

9. Maatouk DM, DiNapoli L, Alvers A, Parker KL, Taketo MM, Capel B. Stabilization of beta-catenin in XY gonads causes male-to-female sex-reversal. Hum Mol Genet. 2008;17:2949–55.

10. Detti L, Martin DC, Williams LJ. Applicability of adult techniques for ovarian preservation to childhood cancer patients. Assist Reprod Genet. 2012 July 21. [Epub ahead of print]

11. Jost A. A new look at the mechanisms controlling sex differentiation in mammals. Johns Hopkins Med J 1972;130:38–53.

12. Wagner T, Wirth J, Meyer J, Zabel B, Held M, Zimmer J, Pasantes J, Bricarelli FD, Keutel J, Hustert E. Autosomal sex reversal and campomelic dysplasia are caused by mutations in and around the SRY-related gene SOX9. Cell. 1994;79:1111–20.

13. Huang B, Wang S, Ning Y, Lamb AN, Bartley J. Autosomal XX sex reversal caused by duplication of SOX9. Am J Med Genet. 1999;87:349–53.

14. Herdt GH, Davidson J. The Sambia 'turnim-man': sociocultural and clinical aspects of gender formation in male pseudohermaphrodites with 5-alpha-reductase deficiency in Papua New Guinea. Arch Sex Behav. 1988;17:33–56.

15. Povey AC, Stocks SJ. Epidemiology and trends in male subfertility. Hum Fertil (Camb). 2010;13:182–8.

16. Braw-Tal R. Endocrine disruptors and timing of human exposure. Pediatr Endocrinol Rev. 2010;8:41–6.

17. Shine R, Peek J, Birdsall M. Declining sperm quality in New Zealand over 20 years. N Z Med J. 2008;121:50–6.

Chapter 13: Wired

1. Sanai N, Nguyen T, Ihrie RA, Mirzadeh Z, Tsai HH, Wong M, Gupta N, Berger MS, Huang E, Garcia-Verdugo JM, Rowitch DH, Alvarez-Buylla A. Corridors of migrating neurons in the human brain and their decline during infancy. Nature. 2011;478:382–6.

2. Lowery LA, Van Vactor D. The trip of the tip: understanding the growth cone machinery. Nat Rev Mol Cell Biol. 2009;10:332–43.

3. Bard L, Boscher C, Lambert M, Mège RM, Choquet D, Thoumine O. A molecular clutch between the actin flow and N-cadherin adhesions drives growth cone migration. J Neurosci. 2008;28:5879–90.

4. Bateman J, Van Vactor D. The Trio family of guanine-nucleotide-exchange factors: regulators of axon guidance. J Cell Sci. 2001;114:1973–80.

5. Davies JA, Cook GM. Growth cone inhibition—an important mechanism in neural development? Bioessays. 1991;13:11–15.

6. Long H, Sabatier C, Ma L, Plump A, Yuan W, Ornitz DM, Tamada A, Murakami F, Goodman CS, Tessier-Lavigne M. Conserved roles for Slit and Robo proteins in midline commissural axon guidance. Neuron. 2004;42:213–23.

7. Parra LM, Zou Y. Sonic hedgehog induces response of commissural axons to Semaphorin repulsion during midline crossing. Nat Neurosci. 2009;13:29–35.

8. Reeber SL, Kaprielian Z. Leaving the midline: how Robo receptors regulate the guidance of post-crossing spinal commissural axons. Cell Adh Migr. 2009;3:300–4.

9. Farmer WT, Altick AL, Nural HF, Dugan JP, Kidd T, Charron F, Mastick GS. Pioneer longitudinal axons navigate using floor plate and Slit/Robo signals. Development. 2008;135:3643–53.

10. Scicolone G, Ortalli AL, Carri NG. Key roles of Ephs and ephrins in retinotectal topographic map formation. Brain Res Bull. 2009;79:227–47.

11. Erskine L, Herrera E. The retinal ganglion cell axon's journey: insights into molecular mechanisms of axon guidance. Dev Biol. 2007;308:1–14.

12. Oster SF, Bodeker MO, He F, Sretavan DW. Invariant Sema5A inhibition serves an ensheathing function during optic nerve development. Development 2003;130:775–84.

13. Wang J, Chan CK, Taylor JS, Chan SO. The growth-inhibitory protein Nogo is involved in midline routing of axons in the mouse optic chiasm. J Neurosci Res. 2008;86:2581–90.

14. Kuwajima T, Yoshida Y, Takegahara N, Petros TJ, Kumanogoh A, Jessell TM, Sakurai T, Mason C. Optic chiasm presentation of Semaphorin6D in the context of Plexin-A1 and Nr-CAM promotes retinal axon midline crossing. Neuron. 2012;74:676–90.

15. Erskine L, Reijntjes S, Pratt T, Denti L, Schwarz Q, Vieira JM, Alakakone B, Shewan D, Ruhrberg C. VEGF signaling through neuropilin 1 guides commissural axon crossing at the optic chiasm Neuron. 2011;70:951–65.

16. Wynshaw-Boris A, Pramparo T, Youn YH, Hirotsune S. Lissencephaly: mechanistic insights from animal models and potential therapeutic strategies. Semin Cell Dev Biol. 2010;21:823–30.

17. Schäfer MK, Altevogt P. L1CAM malfunction in the nervous system and human carcinomas. Cell Mol Life Sci. 2010;67:2425–37.

18. Fransen E, Van Camp G, Vits L, Willems PJ. L1-associated diseases: clinical geneticists divide, molecular geneticists unite. Hum Mol Genet. 1997;6:1625–32.

19. Jen JC, Chan WM, Bosley TM, Wan J, Carr JR, Rüb U, Shattuck D, Salamon G, Kudo LC, Ou J, Lin DD, Salih MA, Kansu T, Al Dhalaan H, Al Zayed Z, MacDonald DB, Stigsby B, Plaitakis A, Dretakis EK, Gottlob I, Pieh C, Traboulsi EI, Wang Q, Wang L, Andrews C, Yamada K, Demer JL, Karim S, Alger JR, Geschwind DH, Deller T, Sicotte NL, Nelson SF, Baloh RW, Engle EC. Mutations in a human ROBO gene disrupt hindbrain axon pathway crossing and morphogenesis. Science. 2004;304:1509–13.

Chapter 14: Dying to be Human

1. Pole RJ, Qi BQ, Beasley SW. Patterns of apoptosis during degeneration of the pronephros and mesonephros. J Urol. 2002;167:269–71.

2. Zuzarte-Luís V, Hurlé JM. Programmed cell death in the developing limb. Int J Dev Biol. 2002;46:871–6.

3. Zakeri Z, Quaglino D, Ahuja HS. Apoptotic cell death in the mouse limb and its suppression in the hammertoe mutant. Dev Biol. 1994;165:294–7.

4. Merino R, Rodriguez-Leon J, Macias D, Gañan Y, Economides AN, Hurle JM. The BMP antagonist Gremlin regulates outgrowth, chondrogenesis and programmed cell death in the developing limb. Development. 1999;126:5515–22.

5. Oppenheim RW. Cell death during development of the nervous system. Annu Rev Neurosci. 1991;14:453–501.

6. Hutchins JB, Barger SW. Why neurons die: cell death in the nervous system. Anat Rec. 1998;253:79–90.

7. Hamburger V. The effects of wing bud extirpation on the development of the central nervous system in chick embryos. J Exp Zool. 1934;68:449–94.

8. Lanser ME, Fallon JF. Development of the lateral motor column in the limbless mutant chick embryo. J Neurosci. 1984;4:2043–50.

9. Lamb AH. Target dependency of developing motoneurons in Xenopus laevis. J Comp Neurol. 1981;203:157–71.

10. Tanaka H, Landmesser LT. Cell death of lumbosacral motoneurons in chick, quail, and chick-quail chimera embryos: a test of the quantitative matching hypothesis of neuronal cell death. J Neurosci. 1986;6:2889–99.

11. Raff MC. Social controls on cell survival and cell death. Nature. 1992;356:397–400.

12. Sharifi N, Gulley JL, Dahut WL. An update on androgen deprivation therapy for prostate cancer. Endocr Relat Cancer. 2010;17:R305–15.

13. Rick FG, Schally AV, Block NL, Nadji M, Szepeshazi K, Zarandi M, Vidaurre I, Perez R, Halmos G, Szalontay L. Antagonists of growth hormone-releasing hormone (GHRH) reduce prostate size in experimental benign prostatic hyperplasia. Proc Natl Acad Sci U S A. 2011;108:3755–60.

14. Kimmick GG, Muss HB. Endocrine therapy in metastatic breast cancer. Cancer Treat Res. 1998;94:231–54.

Chapter 15: Making Your Mind Up

1. Hebb DO. The organization of behavior. 1949;Wiley.

2. Glanzman DL. Associative Learning: Hebbian Flies. Curr Biol 2005;15:R416–419.

3. Xia S, Miyashita T, Fu TF, Lin WY, Wu CL, Pyzocha L, Lin IR, Saitoe M, Tully T, Chiang AS. NMDA receptors mediate olfactory learning and memory in Drosophila. Curr Biol. 2005;15:603–15.

4. Feldman DE, Knudsen EI. An anatomical basis for visual calibration of the auditory space map in the barn owl's midbrain. J Neurosci. 1997;17:6820–37.

5. Brainard MS, Knudsen EI. Sensitive periods for visual calibration of the auditory space map in the barn owl optic tectum. J Neurosci. 1998;18:3929–42.

6. Tomoda A, Sheu YS, Rabi K, Suzuki H, Navalta CP, Polcari A, Teicher MH. (2010) Exposure to parental verbal abuse is associated with increased gray matter volume in superior temporal gyrus. Neuroimage. 2010 May 17.

7. Teicher MH, Samson JA, Sheu YS, Polcari A, McGreenery CE. Hurtful Words: Association of Exposure to Peer Verbal Abuse With Elevated Psychiatric Symptom Scores and Corpus Callosum Abnormalities. 2010;167:1464–71.

8. Tomoda A, Suzuki H, Rabi K, Sheu YS, Polcari A, Teicher MH. Reduced prefrontal cortical gray matter volume in young adults exposed to harsh corporal punishment. Neuroimage. 2009;47Suppl2:T66–71.

9. Tomoda A, Navalta CP, Polcari A, Sadato N, Teicher MH. Childhood sexual abuse is associated with reduced gray matter volume in visual cortex of young women. Biol Psychiatry. 2009;66:642–8.

Chapter 16: A Sense of Proportion

1. Raben MS. Treatment of a pituitary dwarf with human growth hormone. J Clin Endocrinol Metab. 1958;18:901–3.

2. Kemp SF. (2009) Insulin-like growth factor-I deficiency in children with growth hormone insensitivity: current and future treatment options. *BioDrugs.* 2009; **23**: 155–63.

3. Kemp SF, Frindik JP. Emerging options in growth hormone therapy: an update.Drug Des Devel Ther. 2011;**5**:411–19.

4. Giustina A, Mazziotti G, Canalis E. Growth hormone, insulin-like growth factors, and the skeleton. *Endocr Rev.* 2008;**29**:535–59.

5. Arman A, Yüksel B, Coker A, Sarioz O, Temiz F, Topaloglu AK. Novel growth hormone receptor gene mutation in a patient with Laron syndrome. *J Pediatr Endocrinol Metab.* 2010;**23**:407–14.

6. Laron Z. The GH-IGF1 axis and longevity: the paradigm of IGF1 deficiency. *Hormones.* 2008;**7**:24–7.

7. Cawthorne T. Toulouse-Lautrec—triumph over infirmity. *Proc Roy Soc Med* 1970;**63**:800–5.

8. Maroteaux P, Lamy M. The malady of Toulouse-Lautrec. *JAMA.* 1995;**191**:715–17.

9. Maroteaux P. Toulouse-Lautrec's diagnosis. *Nat Genet.* 1995;**11**:362–3.

10. Frey JB. What dwarfed Toulouse-Lautrec? *Nat Genet.* 1995;**10**:128–30.

11. Gelb BD, Shi GP, Chapman HA, Desnick RJ. Pycnodysostosis, a lysosomal disease caused by cathepsin K deficiency. *Science.* 1996;**273**:1236–8.

12. Toral-López J, Gonzalez-Huerta LM, Sosa B, Orozco S, González HP, Cuevas-Covarrubias SA. Familial pycnodysostosis: identification of a novel mutation in the CTSK gene (cathepsin K). *J Investig Med.* 2011;**59**:277–80.

13. Chen W, Yang S, Abe Y, Li M, Wang Y, Shao J, Li E, Li YP. Novel pycnodysostosis mouse model uncovers cathepsin K function as a potential regulator of osteoclast apoptosis and senescence. *Hum Mol Genet.* 2007;**16**:410–23.

14. Boskey AL, Gelb BD, Pourmand E, Kudrashov V, Doty SB, Spevak L, Schaffler MB. Ablation of cathepsin k activity in the young mouse causes hypermineralization of long bone and growth plates. *Calcif Tissue Int.* 2009;**84**:229–39.

15. Rothenbühler A, Piquard C, Gueorguieva I, Lahlou N, Linglart A, Bougnères P. Near normalization of adult height and body proportions by growth hormone in pycnodysostosis. *J Clin Endocrinol Metab.* 2010;**95**:2827–31.

16. Shiang R, Thompson LM, Zhu YZ, Church DM, Fielder TJ, Bocian M, Winokur ST, Wasmuth JJ. Mutations in the transmembrane domain of FGFR3 cause the most common genetic form of dwarfism, achondroplasia. *Cell.* 1994;**78**:335–42.

17. Rousseau F, Bonaventure J, Legeai-Mallet L, Pelet A, Rozet JM, Maroteaux P, Le Merrer M, Munnich A. Mutations in the gene encoding fibroblast growth factor receptor-3 in achondroplasia. *Nature.* 1994;**371**:252–4.

18. Richette P, Bardin T, Stheneur C. Achondroplasia: from genotype to phenotype. *Joint Bone Spine.* 2008;**75**:125–30.

19. Baron J, Klein KO, Colli MJ, Yanovski JA, Novosad JA, Bacher JD, Cutler GB Jr Catch-up growth after glucocorticoid excess: a mechanism intrinsic to the growth plate. *Endocrinology.* 1994;**135**:1367–71.

20. Chagin AS, Karimian E, Sundström K, Eriksson E, Sävendahl L. Catch-up growth after dexamethasone withdrawal occurs in cultured postnatal rat metatarsal bones. *J Endocrinol.* 2010;**204**:21–9.

21. Kronenberg HM. PTHrP and skeletal development. *Ann N Y Acad Sci.* 2006;**1068**:1–13.

22. Gafni RI, Baron J. (2000) Catch-up growth: possible mechanisms. *Pediatr Nephrol.* 2000;**14**: 616–19.

23. Grumbach MM. Mutations in the synthesis and action of estrogen: the critical role in the male of estrogen on pubertal growth, skeletal maturation, and bone mass. *Ann N Y Acad Sci.* 2004;**1038**:7–13.

24. Chagin AS, Sävendahl L. Genes of importance in the hormonal regulation of growth plate cartilage. *Horm Res.* 2009;**71** Suppl 2:41–7.

25. Grumbach MM. Estrogen, bone, growth and sex: a sea change in conventional wisdom. *J Pediatr Endocrinol Metab.* 2000;**13** Suppl 6:1439–55.

26. Eastell R. Role of oestrogen in the regulation of bone turnover at the menarche. *J Endocrinol.* 2005;**185**:223–34.

27. Jones IE, Williams SM, Dow N, Goulding A. How many children remain fracture-free during growth? a longitudinal study of children and adolescents participating in the Dunedin Multidisciplinary Health and Development Study. *Osteoporos Int.* 2002;**13**:990–5.

28. Pietramaggiori G, Liu P, Scherer SS, Kaipainen A, Prsa MJ, Mayer H, Newalder J, Alperovich M, Mentzer SJ, Konerding MA, Huang S, Ingber DE, Orgill DP. Tensile forces stimulate vascular remodeling and epidermal cell proliferation in living skin. *Ann Surg.* 2007 **246**:896–902.

29. Nelson CM, Jean RP, Tan JL, Liu WF, Sniadecki NJ, Spector AA, Chen CS. Emergent patterns of growth controlled by multicellular form and mechanics. *Proc Natl Acad Sci USA.* 2005;**102**:11594–9.

30. Ingber DE. Mechanical control of tissue growth: function follows form. *Proc Natl Acad Sci USA.* 2005;**102**:11571–2.

31. Golde A. Chemical changes in chick embryo cells infected with Rous Sarcoma Virus in vitro. *Virology* 1962;**16**:9–20.

32. Grusche FA, Richardson HE, Harvey KF. Upstream regulation of the hippo size control pathway. *Curr Biol.* 2010;**20**:R574–82.

33. Doggett K, Grusche FA, Richardson HE, Brumby AM. Loss of the Drosophila cell polarity regulator Scribbled promotes epithelial tissue overgrowth and cooperation with oncogenic Ras-Raf through impaired Hippo pathway signaling. *BMC Dev Biol.* 2011;**11**:57.

34. Silber SJ. Growth of baby kidneys transplanted into adults. *Arch Surg* 1976;**111**:75–7.

35. Metcalf D. (1964) Restricted growth capacity of multiple spleen grafts. *Transplantation* 1964;**2**:387–92.

36. Metcalf D. (1963) The autonomous behaviour of normal thymus grafts. *Aust J Exp Biol Med Sci* 1963;**41**:437–47.

Chapter 17: Making Friends and Facing Enemies

1. Xu J and Gordon JI. Honor thy symbionts. Proc Natl Acad Sci U S A. 2003;100: 10452–9.

2. O'Hara AM, Shanahan F. The gut flora as a forgotten organ. EMBO Rep. 2006;7: 688–93.

3. Scharlau D, Borowicki A, Habermann N, Hofmann T, Klenow S, Miene C, Munjal U, Stein K, Glei M. Mechanisms of primary cancer prevention by butyrate and other products formed during gut flora-mediated fermentation of dietary fibre. Mutat Res. 2009 July–August;682(1):39–53.

4. Salaspuro MP. Acetaldehyde, microbes, and cancer of the digestive tract. Crit Rev Clin Lab Sci. 2003;40:183–208.

5. Hill MJ. Intestinal flora and endogenous vitamin synthesis. Eur J Cancer Prev. 1997 March;6 Suppl 1:S43–5.

6. Lazarenko L, Babenko L, Sichel LS, Pidgorskyi V, Mokrozub V, Voronkova O, Spivak M. Antagonistic Action of Lactobacilli and Bifidobacteria in Relation to Staphylococcus aureus and Their Influence on the Immune Response in Cases of Intravaginal Staphylococcosis in Mice. Probiotics Antimicrob Proteins. 2012 June;4(2):78–89.

7. Marbieri M (Editor) Biosemiotics: information, codes and signs in living systems. 2007; Nova, New York.

8. Bry L, Falk PG, Midtvedt T, Gordon JI. A model of host-microbial interactions in an open mammalian ecosystem. Science. 1996 September 6;273(5280):1380–3.

9. Stappenbeck TS, Hooper LV, Gordon JI. Developmental regulation of intestinal angiogenesis by indigenous microbes via Paneth cells.Proc Natl Acad Sci U S A. 2002; 99: 15451–5.

10. Hooper LV, Stappenbeck TS, Hong CV, Gordon JI. Angiogenins: a new class of microbicidal proteins involved in innate immunity. Nat Immunol. 2003 March; 4(3): 269–73.

11. Matzinger P. Tolerance, danger, and the extended family. Annu Rev Immunol. 1994;12:991–1045.

12. Nikolich-Zugich J, Slifka MK, Messaoudi I. The many important facets of T-cell repertoire diversity. Nat Rev Immunol. 2004;4:123–32.

13. Takahama Y, Nitta T, Mat Ripen A, Nitta S, Murata S, Tanaka K. Role of thymic cortex-specific self-peptides in positive selection of T cells. Semin Immunol. 2010 October;22(5):287–93.

14. Davies J, Sheil B, Shanahan F. Bacterial signalling overrides cytokine signalling and modifies dendritic cell differentiation. Immunology. 2009;128:e805–15.

15. Zeuthen LH, Fink LN, Frokiaer H. Epithelial cells prime the immune response to an array of gut-derived commensals towards a tolerogenic phenotype through distinct actions of thymic stromal lymphopoietin and transforming growth factor-beta. Immunology. 2008;123:197–208.

16. Mazmanian SK, Liu CH, Tzianabos AO, Kasper DL. An immunomodulatory molecule of symbiotic bacteria directs maturation of the host immune system. Cell. 2005;122:107–18.

17. Von Hertzen LC, Haahtela T. Asthma and atopy—the price of affluence? Allergy. 2004;59:124–37.

Chapter 18: Maintenance Mode

1. Luisi PL. The emergence of life. Particularly pp 23–6. 2006; Cambridge University Press.

2. Barker N, van de Wetering M, Clevers H. The intestinal stem cell. Genes and Development 2008;22:1856–64.

3. Potten CS, Gandara R, Mahida YR, Loeffler M, Wright NA. The stem cells of small intestinal crypts: where are they? Cell Prolif. 2009;42:731–50.

4. Batlle E, Henderson JT, Beghtel H, van den Born MM, Sancho E, Huls G, Meeldijk J, Robertson J, van de Wetering M, Pawson T, Clevers H. Beta-catenin and TCF mediate

cell positioning in the intestinal epithelium by controlling the expression of EphB/ephrinB. Cell. 2002;111(2):251–63.

5. Neal MD, Richardson WM, Sodhi CP, Russo A, Hackam DJ. Intestinal stem cells and their roles during mucosal injury and repair. Surg Res. 2011;167:1–8.

6. Farin HF, van Es JH, Clevers H. Redundant Sources of Wnt Regulate Intestinal Stem Cells and Promote Formation of Paneth Cells. Gastroenterology. 2012 August 22. [Epub ahead of print]

7. Schuijers J, Clevers H. Adult mammalian stem cells: the role of Wnt, Lgr5 and R-spondins.EMBO J. 2012 May 22;31(12):2685–96.

8. Di Girolamo N. Stem cells of the human cornea. Br Med Bull. 2011;100:191–207.

9. Mort RL, Ramaesh T, Kleinjan DA, Morley SD, West JD. Mosaic analysis of stem cell function and wound healing in the mouse corneal epithelium. BMC Dev Biol. 2009 January 7;9:4.

10. Collinson JM, Morris L, Reid AI, Ramaesh T, Keighren MA, Flockhart JH, Hill RE, Tan SS, Ramaesh K, Dhillon B, West JD. Clonal analysis of patterns of growth, stem cell activity, and cell movement during the development and maintenance of the murine corneal epithelium. Dev Dyn. 2002 August;224(4):432–40.

11. Romagnani P. Toward the identification of a 'renopoietic system'? Stem Cells. 2009 September;27(9):2247–53.

12. Kirouac DC, Madlambayan GJ, Yu M, Sykes EA, Ito C, Zandstra PW. Cell-cell interaction networks regulate blood stem and progenitor cell fate. Mol Syst Biol. 2009;5:293.

13. Abdallah BM, Kassem M. Human mesenchymal stem cells: from basic biology to clinical applications. Gene Ther. 2008;15:109–16.

14. Höcht-Zeisberg E, Kahnert H, Guan K, Wulf G, Hemmerlein B, Schlott T, Tenderich G, Körfer R, Raute-Kreinsen U, Hasenfuss G. Cellular repopulation of myocardial infarction in patients with sex-mismatched heart transplantation. Eur Heart J. 2004;25:749–58.

15. Matsumoto T, Okamoto R, Yajima T, Mori T, Okamoto S, Ikeda Y, Mukai M, Yamazaki M, Oshima S, Tsuchiya K, Nakamura T, Kanai T, Okano H, Inazawa J, Hibi T, Watanabe M. Increase of bone marrow-derived secretory lineage epithelial cells during regeneration in the human intestine. Gastroenterology. 2005;128:1851–67.

16. Brittan M, Hunt T, Jeffery R, Poulsom R, Forbes SJ, Hodivala-Dilke K, Goldman J, Alison MR, Wright NA. Bone marrow derivation of pericryptal myofibroblasts in the mouse and human small intestine and colon. Gut. 2002;50:752–7.

17. Sostak P, Theil D, Stepp H, Roeber S, Kretzschmar HA, Straube A. Detection of bone marrow-derived cells expressing a neural phenotype in the human brain. Neuropathol Exp Neurol. 2007;66:110–16.

18. Crain BJ, Tran SD, Mezey E. Transplanted human bone marrow cells generate new brain cells. J Neurol Sci. 2005;233:121–3.

19. Mezey E, Key S, Vogelsang G, Szalayova I, Lange GD, Crain B. Transplanted bone marrow generates new neurons in human brains. Proc Natl Acad Sci U S A. 2003;100:1364–9.

20. Poulsom R, Forbes SJ, Hodivala-Dilke K, Ryan E, Wyles S, Navaratnarasah S, Jeffery R, Hunt T, Alison M, Cook T, Pusey C, Wright NA. Bone marrow contributes to renal parenchymal turnover and regeneration. J Pathol. 2001;195:229–35.

21. Du H, Taylor HS. Contribution of bone marrow-derived stem cells to endometrium and endometriosis. Stem Cells. 2007;25:2082–6.

22. Ikoma T, Kyo S, Maida Y, Ozaki S, Takakura M, Nakao S, Inoue M. Bone marrow-derived cells from male donors can compose endometrial glands in female transplant recipients.Am J Obstet Gynecol. 2009;201:608.e1–8.

23. O'Donoghue K, Chan J, de la Fuente J, Kennea N, Sandison A, Anderson JR, Roberts IA, Fisk NM. Microchimerism in female bone marrow and bone decades after fetal mesenchymal stem-cell trafficking in pregnancy. Lancet. 2004;364:179–82.

24. Lepez T, Vandewoestyne M, Hussain S, Van Nieuwerburgh F, Poppe K, Velkeniers B, Kaufman JM, Deforce D. Fetal microchimeric cells in blood of women with an auto-immune thyroid disease. PLoS One. 2011;6(12):e29646.

25. Soldini D, Moreno E, Martin V, Gratwohl A, Marone C, Mazzucchelli L. BM-derived cells randomly contribute to neoplastic and non-neoplastic epithelial tissues at low rates. Bone Marrow Transplant. 2008;42:749–55.

26. Bayes-Genis A, Bellosillo B, de la Calle O, Salido M, Roura S, Ristol FS, Soler C, Martinez M, Espinet B, Serrano S, Bayes de Luna A, Cinca J. Identification of male cardiomyocytes of extracardiac origin in the hearts of women with male progeny: male fetal cell microchimerism of the heart. J Heart Lung Transplant. 2005;24:2179–83.

27. Mettler FA, Gus'kova AK, Gusev I. Health effects in those with acute radiation sickness from teh Chernobyl accident. Health Physics. 2007;93:462–9.

28. Somosy Z, Horváth G, Telbisz A, Réz G, Pálfia Z. Morphological aspects of ionizing radiation response of small intestine. Micron. 2002;33(2):167–78.

29. Burgess AW, Faux MC, Layton MJ, Ramsay RG. Wnt signaling and colon tumorigene-sis—a view from the periphery. Exp Cell Res. 2011 November 15;317(19):2748–58.

30. Ricci-Vitiani L, Fabrizi E, Palio E, De Maria R. Colon cancer stem cells. J Mol Med. 2009;87:1097–104.

31. Frosina G. The bright and the dark sides of DNA repair in stem cells. J Biomed Biotechnol. 2010;2010:845396.

32. Frank NY, Schatton T, Frank MH. The therapeutic promise of the cancer stem cell concept. J Clin Invest. 2010;120:41–50.

33. Rosen JM, Jordan CT. The increasing complexity of the cancer stem cell paradigm. Science. 2009;324:1670–3.

34. Lasagni L, Romagnani P. Glomerular epithelial stem cells: the good, the bad, and the ugly. J Am Soc Nephrol. 2010 October;21(10):1612–19.

35. Secker GA, Daniels JT. Corneal epithelial stem cells: deficiency and regulation. Stem Cell Rev. 2008 September;4(3):159–68.

Chapter 19: Perspectives

1. Burger A, Davidson D, Baldock R. Formalization of mouse embryo anatomy. Bioinformatics 2004;20:259–67.

2. Johannsen W (1909) Elemente der Exakten Erblichkeitslehre. 1909; Gustav Fisher, Jena.

3. Wolterek R. Weitere experimentelle Untersuchingen über Artveränderung, speziell über das niden. Versuch. Deutech. Zool. Ges. 1909: 110–72.

4. Kittler R, Buchholz F. RNA interference: gene silencing in the fast lane. Semin Cancer Biol. 2003;13:259–65.

5. Bosher JM, Labouesse M. RNA interference: genetic wand and genetic watchdog. Nat Cell Biol. 2000;2:E31–6.

6. Plasterk RH, Ketting RF. The silence of the genes. Curr Opin Genet Dev 2000; 10:562–7.

7. Mangan S, Alon U. Structure and function of the feed-forward loop network motif. Proc Natl Acad Sci U S A. 2003;100:11980–5.

8. Unbekandt M, Davies JA. Dissociation of embryonic kidneys followed by reaggregation allows the formation of renal tissues. Kidney Int. 2010;77:407–16.

9. Ganeva V, Unbekandt M, Davies JA. An improved kidney dissociation and reaggregation culture system results in nephrons arranged organotypically around a single collecting duct system. Organogenesis. 2011;7:83–7.

10. Macchiarini P, Jungebluth P, Go T, Asnaghi MA, Rees LE, Cogan TA, Dodson A, Martorell J, Bellini S, Parnigotto PP, Dickinson SC, Hollander AP, Mantero S, Conconi MT, Birchall MA. Clinical transplantation of a tissue-engineered airway. Lancet. 2008;372:2023–30.

FURTHER READING

This is a list of books that, like this one, are intended for non-specialists. They will enable readers to explore further some of the topics introduced in the chapters of this book.

Concepts of adaptive self-organization and emergence

Davies JA (2005) Mechanisms of Morphogenesis. San Diego, CA: Elsevier Academic Press. (*This book is intended for an academic audience but Chapter 1.2 is specifically an introduction to adaptive self-organization and emergence in biology and is written at a level similar to this book.*)

Holland JH (1998) Emergence: From chaos to order. Oxford University Press. (*This book has quite a strong computer-science slant.*)

Johnson S (2001) Emergence: The Connected Lives of Ants, Brains, Cities and Software. London: Penguin.

Kelly K (1994) Out of Control: The new biology of machines. London: Fourth Estate. (*This book is very wide ranging—chapter 2, 'Hive Mind', is a superb introduction to adaptive self-organization.*)

Wikipedia article on 'swarm intelligence'.

The inner workings of cells

Kratz RF (2009) Molecular and cell biology for dummies. Hoboken, NJ: Wiley.

Rose S (1999) The chemistry of life. London: Penguin Press Science.

General human anatomy

Baggaley A (Ed) (2001) Human Body. London: Dorling Kindersley. (*Although this book is inexpensive and looks as if it is published for children, it is superb and I find it very useful even for first-year medical students, whose school learning has given them very little knowledge of what lies inside them.*)

Development before birth

Piontelli A (2002) Twins: From fetus to child. London: Routledge.

Sadler TW (2009) Langman's medical embryology. Philadelphia: Lippincott Williams and Wilkins. *(This book is intended for medical students, but while the text is rather dry and concise the diagrams are superb and easily accessible to non-specialists.)*

Wolpert L (2008) The triumph of the embryo. Mineola, NY: Dover.

Human congenital abnormalities

Bondeson J (2006) Freaks: The pig-faced lady of Manchester square and other medical marvels. Stroud, Gloucestershire: NPI media group.

Leroi A (2005) Mutants: on the form, varieties and errors of the human body. New York, NY: Harper Perennial.

Post-natal child development

Meggitt C (2006) Child Development. London: Heinemann. *(This book is aimed mainly at parents.)*

Sex determination

Dreger AD (2000) Hermaphrodites and the medical invention of sex. Cambridge, MA: Harvard.

Jones S (2003) Y: The descent of men. London: Abacus.

Karzakis KA (2008) Fixing sex: Intersex, medical authority and lived experience. Durham, North Carolina: Duke University Press.

The brain and nervous system

Carter R (1998) Mapping the mind. London, UK: Phoenix.

Carter R (2009) The brain book. London: Dorling Kindersley.

Gibb B (2007) The rough guide to the brain. New York: Rough guides.

Greenfield S (1997) The human brain. Phoenix, AZ: Phoenix Mass Market Publications.

Sacks O (2009) The man who mistook his wife for a hat. London: Picador.

Pinker S (2003) The language instinct: The new science of language and mind. London: Penguin Science.

Microbes and defence against them

Crawford DH (2002) The invisible enemy: A natural history of viruses. Oxford: Oxford University Press.

Crawford DH (2009) Deadly companions: How microbes shaped out history. Oxford; New York: Oxford University Press.

Stem cells

Goldstein SB (2010) Stem cells for dummies. Hoboken, NJ: Wiley.

Scott CT (2006) Stem cell now: A brief introduction to the coming medical revolution. New York, NY: Plume.

Weinberg RA (1999) One renegade cell: The quest for the origin of cancer. New York: Basic Books.

Biological communication and codes

Barbieri M (2007) Biosemiotics: Information, codes and signs in living systems. New York: Nova.

Emergence of complexity from simplicity

Buchanan M (2002) Small world. London: Weidenfeld & Nicholson.

Holland JH (1998) Emergence: From chaos to order. Oxford; NewYork: Oxford University Press.

Noble D (2008) The music of life: Biology beyond genes. New York: Oxford University Press.

Insect societies

Hölldobler B, Wilson EO (2009) The super-organism. New York: Norton.

Nature and Nurture

Ridley M (2004) Nature via nurture: Genes, experience and what makes us human. London: Harper Perennial.

SOURCES OF QUOTATIONS AT
HEADS OF CHAPTERS

Almost all of the quotations have taken the famous words of the great and the good and have applied them deliberately out of context because they happen to fit human development rather well. Here are their original contexts.

The possession of knowledge does not kill the sense of wonder and mystery. There is always more mystery.

Scientists often express sentiments like these, but this phrase happens to come from a non-scientist, the French writer, essayist, and diarist Anaïs Nin (1903–1977). The excerpt comes from a paragraph in her diary, about psychoanalysis.

The history of a man for the nine months preceding his birth would probably be far more interesting than all the three-score and ten years that follow it.

In these words, the English poet, critic, essayist, and philosopher Samuel Taylor Coleridge (1772–1834), expresses eloquently the fascination that leads developmental biologists to don their white coats and explore the secrets of life before birth.

I am large, I contain multitudes.

This phrase is quoted from the free-verse poem now called 'Song of Myself', by the American poet and essayist Walt Whitman (1819–1892). The poem appeared in the collection 'Leaves of Grass', which was Whitman's first major poetic work. It was untitled in the first edition, carried the title 'Poem of Walt Whitman, an American' in the 1860 edition, and the simpler title 'Song of Myself' from 1867. In its original context, the line refers to, and perhaps excuses, the ability of a man to hold at the same time mutually contradictory thoughts, feelings, and opinions.

Honest differences are a healthy sign of progress.

These words come from Mohandas Karamchand Gandhi (1869–1948), now more usually called by the honorific titles 'Mahatma' ('Great Soul'), or 'Bapu' ('Father'). In their original context, the words referred to the process of working out a way for different groups of people to live together in peace.

It is not birth, marriage, or death, but gastrulation which is truly the most important time in your life.

This is one of the few quotations in this book that is not taken out of context. Lewis Wolpert (1929–) is one of the UK's leading developmental biologists who has contributed greatly both to original research and to education in the field. He continues to publish interesting and challenging books for both the public and his fellow scientists.

The brain: an apparatus with which we think we think

When Ambrose Bierce (1842–1913) published his Devil's Dictionary, he began a literary trend for writing satirical definitions that shows no sign of abating. Other definitions include 'Lawyer—one skilled in circumvention of the law'.

Nothing is particularly hard if you divide it into small jobs.

Henry Ford (1863–1947) was not the first industrialist to use an assembly line to put together complex machines by moving them along a chain of people, each of whom performed one simple task on every machine. He did, however, do this on such a scale that he both established the Ford Motor Company as America's leading motor manufacture and drove many other industrialists to organize their factories in the same way. The combination of relatively high pay with mind-numbing repetition of the same task over and over again was even called 'Fordism' for a time.

In principio erat verbum ... (In the beginning was the word ...)

These are the opening lines of the Gospel of John, as set out in the Vulgate bible, a translation of existing texts into Latin mainly by Sophronius Eusebius Hieronymus (347–420), later known as St Jerome.

If the path be beautiful, let us not ask where it leads.

This phrase, by the French writer Anatole France, born as Francoise-Anatole Thibault (1844–1924), was chosen for Chapter 8 to emphasize the fact that cells respond to the cues around them, and not to some idea of final destination.

Man, an ingenious assembly of portable plumbing

This definition was offered by the American journalist, novelist, and poet Christopher Morley (1890–1957). The phrase appears in his book 'Human Being' (1931).

It's organ, organ all the time ...

Mrs Morgan uses this phrase to voice her frustration at her husband's obsession on two separate occasions in the radio play 'Under Milk Wood' by Dylan Thomas (1914–1953). Mr Morgan spends a lot of time playing the music of Bach and Palestrina, which might be the reason for his wife's good-natured complaint.

A simple child//that lightly draws its breath//and feels its life in every limb ...

These are the opening lines of the poem 'We are Seven', by William Wordsworth (1770–1850). The meaning is darker than the short quoted excerpt implies: the verse ends with the line 'What should it know of death?' and the following verses describe a young girl who sings at the graves of her two lost siblings.

The reproduction of mankind is a great marvel and mystery. Had God consulted me in the matter, I should have advised him to continue the generation of the species by fashioning them out of clay.

Martin Luther (1483–1546) was a priest and theologian who laid the foundations for Protestantism. This quotation has been read in two distinct ways, either as a wish that the complications of sex did not exist, or as an admission that Luther would never have had the imagination to invent something so wonderful. The sentence that follows it makes a similar statement about the alternation of day and night, and has a similar ambiguity. The quotation comes from the 'On marriage and celibacy' section of 'A commentary on St Paul's epistle to the Galatians' (Luther wrote against the enforced celibacy of the priesthood, and was himself married to the one-time nun, Katherina von Bora).

Only Connect!... Live in fragments no longer.

Connection is one of the central themes of the novel *Howards End*, by E. M. Forster (1879–1970). The complete quotation is *Only connect! That was the whole of her sermon. Only connect the prose and the passion and both will be exalted, and human love will be seen at its height. Live in fragments no longer.*

In the myddest of lyfe we be in death

This phrase appears in the funeral service section of the first edition of the Anglican Book of Common Prayer (1549), generally assumed to have been written by Archbishop Thomas Cranmer (1489–1556). In Common Worship, the successor to the Book of Common Prayer, the phrase is rendered as *'In the midst of life we are in death'*.

Systems—systems—systems—you can't escape them because nature is systematic, and man is a natural phenomenon, and his intelligence is a natural phenomenon.

Donald Crowhurst (1932–1969) was an inventor, businessman, and sailor who entered the first single-handed round-the-world yacht race. Prevented from entering the Southern Ocean by a boat that was too fragile to survive it, and prevented from turning back by the financial consequences failure would have on his family, Crowhurst spent months sailing the South Atlantic utterly alone and in radio silence. During this period, he filled his ship's log with theological and philosophical writings, some thought-provoking, some hauntingly beautiful, and some almost too painful to read. The writing was recovered when his boat was spotted and picked up by a passing ship, her lonely, courageous skipper no longer aboard.

'You're the worst kind; you're high maintenance but you think you're low maintenance'.

In Nora Ephron's script for the hit film *When Harry Met Sally*, Harry Burns speaks these words to Sally Albright during one of their many arguments about the differences between men and women.

Know thyself.

The phrase 'know thyself' was inscribed in the Temple of Apollo at Delphi. It is commonly attributed to the Athenian philosopher Socrates (469–399 BCE) although, as with almost everything about that thinker's life, it is unclear whether this attribution reflects historical fact about Socrates the man, or the habit of later authors, such as Plato, to create a mythical 'Socrates' credited with being the source of almost all ancient wisdom.

INDEX